The [illegible]
of the British Isles

General Editors:
Eric H. Brown and Keith Clayton

Eastern and Central England

In the same series

Scotland *J.B. Sissons*

Northern England *Cuchlaine A.M. King*

Ireland *G.L. Herries Davies and Nicholas Stephens*

In preparation

Wales and Southwest England
Eric H. Brown, D.Q. Bowen and R.S. Waters

Southeast and Southern England *D.K.C. Jones*

Eastern and Central England

Allan Straw
& Keith M. Clayton

Methuen
LONDON AND NEW YORK

First published in 1979
by Methuen & Co. Ltd
11 New Fetter Lane, London EC4P 4EE
Published in the USA by
Methuen & Co.
an associate company of Methuen, Inc.
733 Third Avenue, New York, NY 10017

Printed in Great Britain
at the University Press
Cambridge

ISBN 0 416 84660 2 (hardbound)
ISBN 0 416 84670 X (paperback)

British Library Cataloguing in Publication Data

Straw, Allan
Eastern and central England – (Geomorphology of the British Isles).
1. Landforms – England – Midlands
2. Landforms – England
I. Title II. Clayton, Keith Martin
III. Series
551.4'09424 GB436.E5 79 – 40804

ISBN 0-416-84660-2
ISBN 0-416-84670-X Pbk

Typeset by Red Lion Setters, Holborn, London
Printed in Great Britain
at the University Press
Cambridge

Contents

General editors' preface

The British Isles is a most varied and complex part of the world, and over a long period it has been studied in great detail by geologists and geomorphologists. Several general studies of the geology and scenery type have appeared, but despite the wealth of the published literature, more detailed work has been limited to a few regions. Indeed although the most ambitious work, Wooldridge and Linton on Southeast England was first published in 1938, no comparable volume has covered any other part of England.

The main difficulty facing anyone writing about the landforms of the British Isles and their history is the paradox that despite a very considerable literature, the gaps and uncertainties seem to diminish very little over the years. It is often said that each piece of research exposes half a dozen new problems, and so far most of the published literature has tended to add to the complexity of our knowledge. The only way out of this problem is through the establishment of more general concepts into which this detail will fit, and this is the role of this series of books on the different parts of the British Isles. The synthesis is certainly not easy, and it is obviously incomplete. Our understanding of the present land surface and its relationship with the underlying geology (whether 'solid' or 'drift') is relatively secure, but the historical background to the present, complicated as it is by the fluctuating environments of Quaternary time, is less well known. Each author has brought his own special interests and expertise to the synthesis of the region he knows best. To conclude the series, a summary volume by the editors will seek common strands in the regional summaries and erect a national framework within which the more detailed regional accounts will fit.

While no standard approach has been imposed on the contributing authors, care has been taken to be consistent in the use of terminology and the recently elaborated Quaternary time-scale (Mitchell *et al.* 1973) has everywhere been used. Correlation tables are provided for each region, while the maps and diagrams will be found particularly valuable since those in the published literature are so scattered, and generally refer to quite small areas. Above all, these volumes will make the existing literature accessible to the greater number who seek a wider evolutionary understanding of the landforms of the very diverse regions of the British Isles.

Acknowledgements

Grateful acknowledgment is made generally to numerous colleagues and other friends who have offered constructive criticism and support, frequently in the field, of many of the ideas and interpretations presented in the text, and in particular to the late Professor D.L. Linton for his initial encouragement and guidance.

Sincere thanks are offered to Mr R.L. Fry, Chief Cartographer in the Department of Geography, University of Exeter, for the skill and care taken in drawing the illustrations.

Allan Straw

To Beryl,
for patience and encouragement

PART I

Eastern England

Allan Straw

1 The geological background

Introduction

Eastern England can be defined loosely as that part of England east of the Pennines and the Midlands between northeast England and southeast England. Though the boundaries adopted make it a region of convenience (fig. 1.1), eastern England has an identity and includes the historic 'pays' of East Anglia, the Fens, Sherwood, Lindsey, Holderness, Cleveland and Teesside in their entirety. Geologically, the outcropping rocks are sedimentary and almost all of Mesozoic age. Physiographically, with the exception of Durham, it is the northern portion of lowland scarpland England, the constituent cuestas being transected by many gaps, broad as the Wash, narrow as at Lincoln, estuarine as the Humber and dry as at Ancaster. Structurally it is part of the raised western rim of the Mesozoic North Sea basin, so that the regional dips are mainly eastward. Geomorphologically the land form is the result of Tertiary and Quaternary denudation and glaciation, for the chalk strata which once covered the whole region have been pared down and back to east Yorkshire, east Lincolnshire and East Anglia, and the once-enveloping ice sheets have left their erosional as well as depositional marks. The landscape is therefore polygenetic, but it must be appreciated from the outset that because each new set of processes was constrained to operate on predetermined features, continuous rather than episodic evolution has been the pattern, and the landscape is as dynamic today as it has been in the past.

Within the region steep slopes and high land occur only in northeast Yorkshire. Only 10% of the area is higher than 200 m, and some 65% lies below 50 m. Most of the region is really low-lying and the surface forms are subtle rather than bold, but the geomorphic mix is as complex as in any other part of the British Isles.

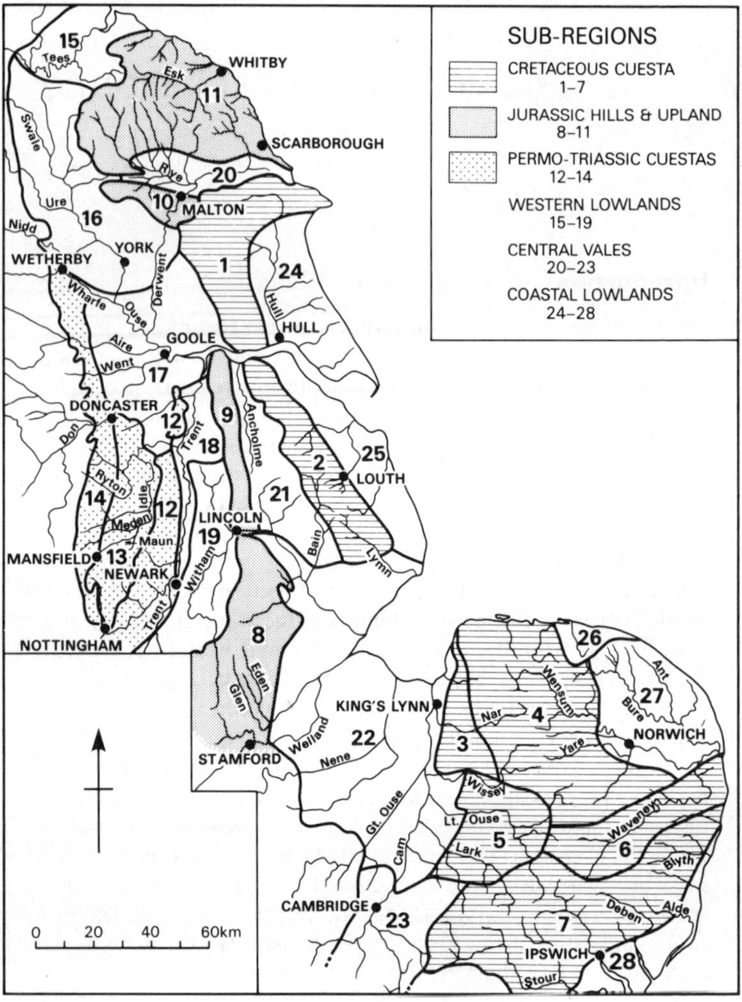

Fig. 1.1 Drainage systems and sub-regions of eastern England. Numbers refer to areas described in the text.

Geological and structural bases

Eastern England comprises the unsubmerged part of the western margin of a major epicontinental sedimentary basin that extends across northern Europe. The present shape of the North Sea is predominantly a consequence of Tertiary subsidence, and Tertiary deposits of very variable lithology extend widely beneath its floor. These reach a maximum thickness of over 3000 m along a north-south axis which lies a little to the east of the central line of the North Sea. Subsidence of the southeastern North Sea has continued into modern times, so that Quaternary sediments over 600 m thick underlie the Low Countries. In England, eastern East Anglia has been directly affected by late Tertiary and Quaternary subsidence, and marine and estuarine deposits of the Crags occur in east Suffolk and east Norfolk. North from East Anglia, Quaternary glacigenic sediments discontinuously overlie the Tertiary and pre-Tertiary strata.

Depression of the North Sea basin probably commenced in late Carboniferous times, and the rocks of eastern England comprise a series of clays, sandstones and limestones with some evaporites that accumulated within it during the Mesozoic era. Deposition was controlled throughout this era by epeirogenic movements which at times determined an oceanic connection for the basin to the north, at others to the south. Alternating marine transgressions and regressions have produced frequent unconformities, condensed deposits and facies changes which render the details of geological history difficult to elucidate. Up to 5000 m of neritic, deltaic, lagoonal and lacustrine sediments have accumulated within the North Sea basin since Permian times, and greater cumulative subsidence toward the centre of the basin has imparted general easterly dips to the rocks, though in northeast Yorkshire, the structural situation is more complex, indicating a more complicated depositional and tectonic history (fig. 1.2).

The geological history of eastern England is not the main burden of this essay, and readers are referred to other texts (Kent 1967, 1975; Sylvester-Bradley and Ford 1968; Rayner and Hemingway 1974) but, because the present physiography of eastern England is in part a response to rock type, some aspects of geological and particularly structural history must be mentioned.

Three primary structural units provide the foundation of eastern

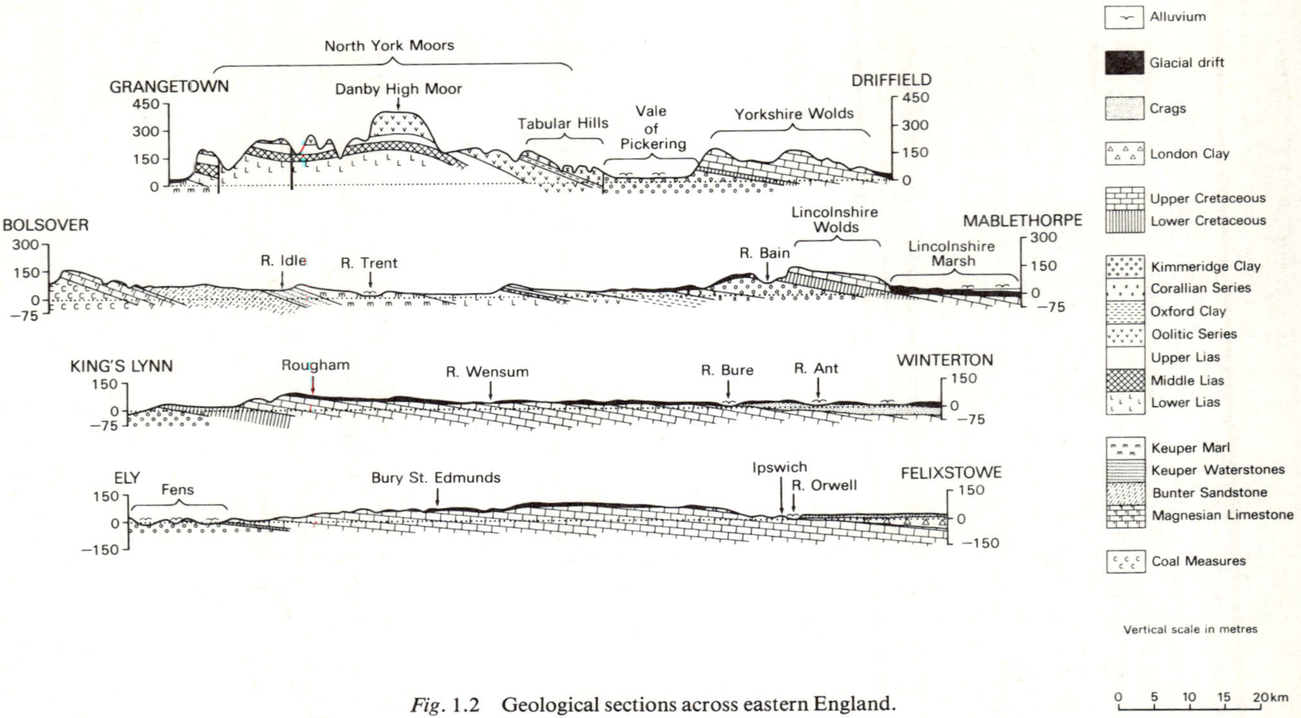

Fig. 1.2 Geological sections across eastern England.

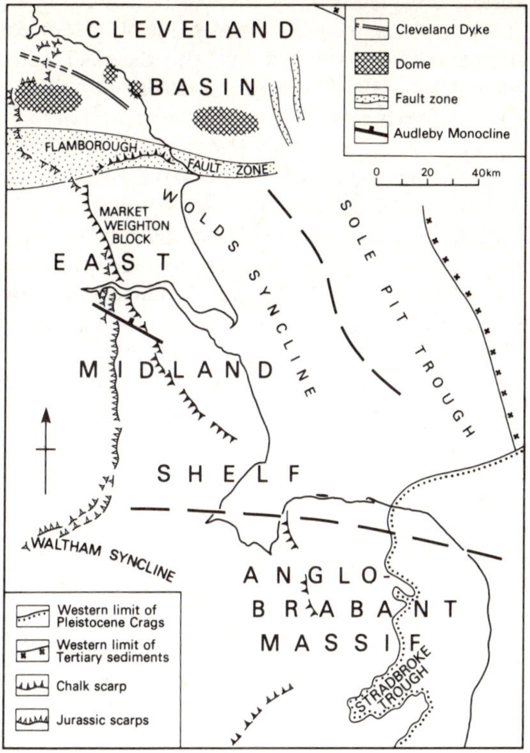

Fig. 1.3 Structural elements of eastern England and adjacent areas of the North Sea. (Partly after Kent, 1967, 1975 and Dingle, 1971.)

England (fig. 1.3). Beneath Suffolk, most of Norfolk and the Fens, and stretching from Wales to the continent lies the concealed *Anglo-Brabant massif*. This ancient block of Lower Palaeozoic rocks is overlain by Jurassic and Cretaceous strata which in Suffolk and Norfolk now dip gently southeast and east at very low angles with a minimum of flexuring and faulting. The broad curvature of the chalk scarp through East Anglia into Lincolnshire suggests a post-Cretaceous arching related to slight uplift of the block.

The chalk of Lincolnshire and east Yorkshire dips seaward into the broad but shallow Tertiary Wolds syncline (Donovan and Dingle

1965), the nose of which controls the eastward sweep of the chalk scarp in east Yorkshire. West of the chalk, the older Mesozoic rocks of Lincolnshire also dip seaward and with the chalk constitute a distinct unit that Kent (1967) believes to coincide with a rigid shelf-like mass of Carboniferous rocks that extends the northern flank of the Anglo-Brabant massif. The eastern limit of this *East Midland shelf* may register with the eastern flank of the Wolds syncline against the complex folded area of the Sole Pit trough (Brunstrom and Walmsley 1969). In the north, the Market Weighton structure (Kent 1955, 1974) lies close to the shelf limit which is determined more precisely by the group of faults, Jurassic through Tertiary in age, that trends eastward from the Howardian Hills. This fault zone may continue offshore into a disturbed belt termed the Flamborough fault zone (Dingle 1971) which leads toward the northwestern end of Sole Pit trough. Westward, the Coxwold-Gilling faults cross the Vale of York to affect the Permian limestones near Ripon.

The third unit differs markedly from the shelf and massif areas in structural style and in the thickness of Jurassic sediments present. Whereas Jurassic sedimentation over the shelf was thin and shallow-water, northward of the faulted 'hinge-line' a deep basin developed along a west-east axis beneath central Cleveland into which over 1000 m of sediments were deposited. Much later, probably in mid-Tertiary times, the *Cleveland basin* rocks were deformed into dome and basin structures with low dips (figs 1.2, 1.3). Landward, the Cleveland dome is responsible for the general northerly and southerly dips off the North York Moors, a pattern complicated by two subsidiary domes centred on the lower Esk valley and Robin Hood's Bay. Other small domes lie seaward off the coast between Whitby and the Tees (Dingle 1971). East and northeast of the Moors, the Cleveland basin continues offshore, still characterized by Tertiary dome and basin structures and minimal faulting. This submerged portion, however, is differentiated from the landward part by a series of *en echelon* faults that trend northnorthwest to southsoutheast between Whitby and Filey. The downthrow of the faults is mainly to the east and rarely more than 60 m, but large transcurrent displacements have been suggested for the Peak fault (Hemingway 1974). As presumed mid-Tertiary structures, they may well have exerted and apparently still exert a control on the location and alignment of this section of the Yorkshire coastline.

Although these three structural units display some contrasts in depositional history, especially in Jurassic times, more uniform deposition occurred at other periods. As a sequel to late-Hercynian epeirogenic movements the Permian sea spread west and south across the eroded Carboniferous platform of eastern England, until its shoreline lay ultimately along the eastern flank of the Pennines and the northern side of the Anglo-Brabant massif. Today this transgression is marked by the angular unconformity (generally about 5°) between the base of the Magnesian Limestone and the Carboniferous rocks and is reflected physiographically by the contraposition of relief trends east and west of the Permian escarpment along much of the Pennine flank.

Another major unconformity lies at the base of the Upper Cretaceous rocks, for the Cenomanian (Lower Chalk) sea was strongly transgressive across much of Britain. Following gentle early and mid-Tertiary movements which warped the chalk strata, a late-Tertiary transgression affected parts of eastern England. Combining with eustatic movements an aberration in the overall Tertiary tilting of eastern England, whereby eastern subsidence and western elevation were somewhat out of phase, could perhaps explain the succession of partial submergence in Pliocene times by early Pleistocene emergence. The Pliocene and Pleistocene Crags of East Anglia rest on an eroded and channelled surface of chalk. In offshore regions, undeformed Tertiary deposits lie east of Sole Pit trough, and northward their western margin roughly parallels the present coast about 80 km away (fig. 1.3). This Pliocene transgression may have been extensive enough to have left some traces in the land form beyond the limits of the Tertiary deposits.

Tectonic structures which have direct physiographic expression are relatively few. Faults across the chalk cuesta in the southern and northern Lincolnshire Wolds, at the Humber and in the northern Yorkshire Wolds, and in the Jurassic areas of south-west Lincolnshire particularly the Howardian Hills and influence valley, and locally scarp, alignments. Fault-line bluffs occur on the north side of the Vale of Pickering. Only the Cleveland domes, which exert marked control on the alignment of cuestas in the North York Moors, exemplify large-scale effects of folding.

2 Regional physiography

The scarplands of eastern England derive from the long-continued erosion of gently inclined rocks of variable lithic character, resulting in clay vales separated by limestone and sandstone cuestas. Viewed broadly the ridge and vale pattern has a deceptive simplicity, complemented by apparently classic drainage systems where the Yorkshire Ouse and the Trent, and the Great Ouse and the Ancholme-lower Witham stream drain strike vales through major water-gaps, and where abandoned cols and abrupt changes of stream course offer spurious proof of river capture. Reconstructions of the stream pattern and land form of eastern England in terms of the Davisian model (Davis 1895; Reed 1901; Versey 1942) retain a superficial attractiveness (Peel and Palmer 1955; de Boer 1974) but fail to explain the many exceptions and ignore the profound consequences of glaciation so clearly indicated by Lewis (1894), Kendall (1902), and Harmer (1910, 1928).

Three tracts of hill country, all basically of cuesta form, though with much variation, constitute the physiographic framework of eastern England (Fig. 1.1). The most continuous cuesta is the Cretaceous rocks in the east. In plan it describes an ogee curve from Flamborough Head through Lincolnshire to the Chilterns. In the centre, Middle Jurassic rocks produce a relatively simple cuesta over the East Midland shelf, but rise to a complex upland over the Cleveland basin in northeast Yorkshire. Longitudinal continuity is prohibited by the overstep of the chalk onto Lower Lias for 26 km north of the Humber. Westward, Triassic and Permian rocks combine to produce a series of low cuestas south of the Don valley, though northward only the Permian cuesta remains prominent.

These hill zones are separated by broad vales opened out on

Triassic, Jurassic and Cretaceous marls and clays. Coastal lowlands are also found east of the chalk where glacial and alluvial sediments are so widespread and continuous that south of Flamborough Head the chalk is only fleetingly exposed in low cliffs in the Humber estuary, and at Hunstanton and Weybourne in Norfolk. The cuestas are transected frequently by river gaps which in several cases serve to define physiographic subregions of individual character (fig. 1.1). In the Tees, Humber and Wash estuaries the sea, now restrained by embankments, still reaches into the inner vales.

The Cretaceous cuesta

Between Flamborough Head and the Humber, the curvilinear *Yorkshire Wolds* (1) are built almost entirely of chalk which carries very little drift and attains a maximum thickness in Yorkshire of 420 m. In the bold north-facing scarp, it overlies the soft grey shales of the Speeton Clay (Lower Cretaceous), which extends a few kilometres from the western end of the scarp beneath the drifts of the Vale of Pickering (Neale 1974). The northwestern scarp, scallopped by four large re-entrants, crosses the Jurassic rocks of the Howardian Hills, but south of Leavening, an impressive, intricately fretted scarp faces westsouthwest, and the Wolds here have a relatively simple cuesta form (Lewin 1969). The central part of the Wolds is a high dissected plateau that declines gradually eastsoutheast from about 240 m O.D. in the west to about 135 m O.D. in the east. It is drained mainly by one deeply incised and excessively branched valley, but northeastward, much of the ground slopes toward the broad and mature Great Wold valley.

South of the Humber the Cretaceous rocks, dipping eastnortheast at about 1:70 into the Wold syncline, have been fashioned into the relatively narrow, rectangular *Lincolnshire Wolds* (2) (Straw 1961a). The northern Wolds display a simple cuesta form where the chalk rests directly on Kimmeridge Clay, which near Elsham incorporates a distinctive local sandstone. The central tract is a dissected plateau at 120-170 m O.D., scarped on the west, but also bluffed steeply along its eastern margin. Deep valleys expose Lower Cretaceous inliers, and west of Louth broad interfluves carry a veneer of glacial deposits. In the southern Wolds, distinct chalk scarps lie only eastward of the Bain and Lymn valleys which are so deeply excavated in Lower Cretaceous rocks that the basal Spilsby

Sandstone is at the surface over wide areas. The scarp is broken decisively by the Bain valley along a section complicated by a thick accumulation of glacial drift (Straw 1966).

South of the Wash gap, Cretaceous rocks dip at less than 1:150 under the whole of East Anglia. In *West Norfolk* (3), Lower Cretaceous rocks crop out along a tapering zone of country between the river Wissey and Hunstanton. The relatively thick basal Sandringham Sands form low, flattish-topped hills, which are frequently heathy or wooded and do not exceed 52 m O.D., between a series of west-draining valleys. Gault clays, overlain by a variety of superficial deposits, underlie a lower strike depression to the east.

Above the Gault depression, a weak chalk scarp, scored by relatively steep-sided valleys, gives way to gently rolling country that reaches 95 m O.D. only around Massingham and Castle Acre. 'Wold' country is poorly developed however and eastward of the crest *central 'High' Norfolk* (4) retains an extensive sheet of highly calcareous till, and chalk is uncovered only in the deeper valleys. Glacifluvial gravels overlie the till in many places, as for example around Syderstone, north of East Dereham, and along the Yare valley, the Wensum and some of its tributaries (Straw 1973).

Within the *Brecklands* (5) thin sands and some gravels overlying chalk and discontinuous glacial drifts so influence surface hydrology and vegetation that the area, which is below 60 m O.D., contrasts sharply with adjacent regions. The sand is residual in part from the weathering of tills, some is glacifluvial and some wind-transported.

East of the Brecklands the *basin of the river Waveney* (6) occupies the eastern part of an ill-defined depression (the Lopham gap) across the chalk area. The Waveney, flowing eastnortheast, has excavated a steep-sided, flat-floored and somewhat sinuous valley through till toward Lowestoft. Westward of Diss the valley merges with that of the Little Ouse river. Both rivers now rise in the peat of Redgrave Fen at about 25 m O.D., but flow in opposite directions.

Over *central 'High' Suffolk* (7) the chalk is mostly concealed by glacial drift, but the ground is generally higher than in Norfolk, reaching 125 m O.D. at Rede and near Balsham in the west and declining southeast in response to the overall gradient of the bedrock surface which is developed across chalk, Tertiary London Clay, Reading and Thanet Beds and, northeast of the Stour, late Pliocene and early Pleistocene sediments (Crags). Most of the Suffolk rivers flow generally to the southeast or east, some sections according with drift-filled troughs or 'buried valleys' (Boswell 1913).

The Jurassic hills and upland

Northward through west Lincolnshire and Humberside, before they disappear beneath the chalk, the Middle Jurassic rocks determine a tapering cuesta, broken by three important gaps at Ancaster, Lincoln and the Humber.

The main scarp-forming rock is the Lincolnshire Limestone (Inferior Oolite) some 33 m thick near Grantham, but thinning to less than 20 m north of Lincoln (Swinnerton and Kent 1976). The western scarp, locally termed 'the Cliff', curves southwest from Grantham, and is fronted by an anterior scarp formed by the Middle Lias Marlstone, a ferruginous oolitic rock some 13 m thick. The *Kesteven plateau* (8) retains extensive spreads of glacial drift, and curiously parallel valleys, eroded through the limestone, carry drainage north and south. Its eastern margin descends abruptly to the Fen margin. The Ancaster gap is a 15 km through-valley once occupied by the river Trent. Northward as the dip increases slightly a more distinct cuesta form prevails, the limestone is more intact and dry valleys are common. The Lincoln gap is shorter, but also has a complex history relating to the Trent (Straw 1963a). Between it and the Humber the cuesta, here known as *Lincoln Edge* (9), is rarely more than 5 km wide or higher than 70 m O.D., and north of Kirton-in-Lindsey is frequently blanketed by Cover Sands. North and south of Scunthorpe thin Lower Lias limestones control a second low cuesta between Scotter and the Humber. North of the estuary, the Jurassic rocks form a narrow bench only along the foot of the chalk scarp as far as Pocklington.

When Middle Jurassic rocks, as the Deltaic Series, emerge from the northwest corner of the Yorkshire Wolds, they combine with the Corallian rocks (Upper Jurassic) to form the double cuesta of the *Howardian Hills* (10). The rocks here dip northeast with outcrops offset by numerous intersecting west-east faults. The Corallian scarp is the more continuous, but the hills rise to 175 m O.D. only in the northwest. They are deeply breached by the Derwent gorge below Malton and terminate sharply in the north against the fault-bounded Coxwold-Gilling gap.

Across the Coxwold-Gilling gap the Jurassic rocks expand

dramatically into 1000 m of lithologically variable sediments within the Cleveland basin. Tertiary uplift and subsequent erosion have created an area of upland, hill and vale, the *North York Moors* (11), where adjustment of land form and structure is close (Palmer 1973). The northern two-thirds of the Moors are underlain by some 250 m of Deltaic Series, incorporating resistant sandstones, such as the highly quartzose fluviatile Moor grit. Northward, Lower Deltaic sandstones are responsible for the bold scarp of the Cleveland Hills, but south of Northallerton, they combine with the younger Corallian rocks to form the more complex west-facing scarp of the Hambleton Hills. The Corallian rocks, 100 m of marine calcareous grits and limestones, extend east across the southern third of the Moors to the coast at Scarborough and introduce a distinctive profile into the landscape, aptly named the Tabular Hills. Their cuesta comprises nine north-pointing salients (four of which rise to or above 300 m O.D.) separated by steep-sided valleys cut so deeply as to expose inliers of Lias. In the east, the intricate ramifications of Troutsdale and Hacknessdale enhance rather than inhibit the 'tabular' character of the hills. On the north flank of the Moors glacial drifts spread 4-10 km inland over a dissected coastal platform, much of which lies between 150 and 215 m O.D. From Whitby to Scarborough the coastal rocks are disturbed by the Peak and Scarborough faults of the Whitby-Filey fracture zone, and several formerly independent streams have been glacially integrated into the Derwent system. The faults together with the contrasting lithologies of the Middle and Lower Jurassic rocks produce great variety of coastal landform under wave attack.

The Permo-Triassic cuestas

Between the Trent at Nottingham and the Don, Triassic rocks in north Nottinghamshire combine with the Permian Magnesian Limestone to produce a hilly tract of redlands comprising three broad cuestas. These are described in more detail in the second half of this book (p.000).

The eastern *Keuper Marl cuesta* (12) emerges hesitatingly from beneath alluvium at Crowle and in the 'Isle of Axholme', and develops southward reaching a maximum height of 145 m O.D. at Loath Hill west of Southwell.

The *Bunter Sandstone* (13), mainly of Pebble Bed facies, thickens north from about 60 m at Nottingham to over 200 m nearer Worksop, but rarely can an escarpment be detected. From its impressive river cliffs in Nottingham to its feeble gravel-covered mounds and ridges near Doncaster, the Bunter provides a contrasting belt of country, the southern two-thirds of which constituted much of the former Sherwood Forest. The surface is dry and the land form is reminiscent of chalk country.

The *Magnesian Limestone cuesta* (14) also trends north from Nottingham, initially as a narrow belt, but around Mansfield it underlies part of the plateau surface that extends from the Bunter sandstone westward on to the Coal Measures. North of Bolsover the strongly dolomitic Lower Magnesian Limestone, 45-90 m thick, occasionally caps a bold west-facing scarp, and about 35 m of Upper Magnesian Limestone occupy part of the outcrop north of Worksop. From the Don valley to Wetherby, the Permian rocks give rise to a narrow zone of dryish country until finally obscured by glacial drift.

The western lowland

From Teesside south to the Jurassic scarp southwest of Grantham a series of broad vales, linked by the Ouse and Trent river systems, displays an unusual degree of geological and geomorphological diversity.

The *lower Tees lowland* (15) lies between the northwestern scarp of the North York Moors and the edge of the Durham plateau between Hartlepool and Darlington. The Triassic foundation is completely overlain by glacial tills, sands and gravels. The Tees enters this lowland in the west and meanders to the estuary at Middlesborough.

The country south of the ill-defined Swale-Tees divides at about 50-60 m O.D. as far as the Escrick moraine constitutes the *Vale of Mowbray* and the larger portion of the traditional *Vale of York* (16). The moraine provides a convenient physiographic boundary, for it separates effectively the northern part of the vale with constructional glacial landforms from that to the south without. The vale coincides with the outcrops of the Bunter Sandstone and Keuper Marl, each over 300 m in thickness, now obscured by glacial and glacifluvial deposits and by lacustrine and alluvial clays and silts particularly along the lower Nidd and Ure and around their confluences with the Ouse. The only peaty area, Snape Mires, lies south of Bedale, half enclosed by a low crescentic morainic ridge. Along the slightly higher flanks glacial deposits furnish a micro-relief of morainic ridges and hummocks, and while they rest mainly on Lower Lias shales in the east they pass onto Permian limestones in the west where, between Boroughbridge, Knaresborough, Tadcaster and York, the surface is a complex of low morainic features associated with poorly integrated drainage.

Within eastern England the alluvial tract of the *Humberhead levels* (17) and their extensions along convergent rivers is second in area only to the Fens. The alluvium is mostly fine-grained, and peat is extensive in the southern part over Hatfield Chase and Thorne Waste. Low, sinuous, sandy roddons marking former stream courses provide slight relief especially over the reclaimed areas of Thorne Waste and the land southeast of Goole. Older lacustrine clays, silts and sands are common north of Goole, but their surface is also virtually featureless and rarely rises above 10 m O.D. East of the river Ouse, thin blown sands derived from glacifluvial, lacustrine and fluvial sediments provide a contrasting surface in terms of soils and land use over wide areas (Matthews 1970). Keuper Marl remains concealed except in a hill near Market Weighton, and the Bunter appears only southwest of Selby, where two curious knolls reaching 50 m O.D. break the uniformity of the surface.

The *lower Trent vale* (18) is relatively narrow, for from Newark to Gainsborough, where it is constricted between the low scarp of limestone-capped Rhaetic shales and the east end of the Barrow Hills, it flows intriguingly along the lower backslope of the Keuper Marl. Southeast of Newark, and within the curve of the Jurassic scarp, is the enigmatic *Vale of Belvoir* (19) opened out not on Triassic rocks but on Lower Lias shales. Between Newark and Lincoln, marking former courses of the Trent, vast spreads of fluviatile gravel survive on and between low discontinuous ridges of thin limestones which continue south over the vale. These are crossed discordantly by the puny river Devon and its tributaries. East of Newark the singular river Witham breaks from the western end of the Ancaster gap, turns northnortheast, collects the rivers Brant and Till, and disappears through the Lincoln gap in search of the Wash.

The central vales

Two widely separated vales, contrasting in extent and alignment, complement the Cretaceous scarps of the northern Yorkshire Wolds and of Lincolnshire and East Anglia.

The uppermost Jurassic rock of the Cleveland Basin is the Kimmeridge Clay. Its outcrop, faulted against the Corallian to the north, and that of the Lower Cretaceous Speeton Clay have been eroded differentially to produce the west-east *Vale of Pickering* (20). The clays are rarely seen, however, for the flat floor below 40 m O.D. is mostly overspread by fluvial and lacustrine sediments and in places by peats. Vale margin deposits are sandy and gravelly, glacial in the east, glacifluvial in the northeast and around Pickering, and lacustrine, aeolian and soliflual along the foot of the chalk escarpment. The North Sea end is now sealed by thick morainic drifts so that the whole vale is drained by the west-flowing Derwent, which both enters and leaves the vale dramatically through fine meltwater channels.

Beneath the Yorkshire Wolds the Kimmeridge Clay is absent over the Market Weighton block but it reappears near the Humber. South from this estuary into Cambridgeshire, all the Upper Jurassic rocks have been gouged out to produce a large strike depression between the Middle Jurassic and Cretaceous cuestas.

The *central Lincolnshire vale* (21) widens south of the Humber and is drained to north and south by way of the Ancholme and Langworth-Witham rivers respectively. The eastern tributaries of the Ancholme have cut through glacial tills into the Jurassic clays to create a low cuesta-like ridge, facing the chalk scarp, between Holton-le-Moor and Wrawby. Here, and southward to Market Rasen, Cover Sands are common. Around Ragby the till sheet is largely intact, but around Tattershall it is concealed by fluvial sands and gravels deposited by the river Bain draining south out of the Wolds.

The Fens (22) are the largest continuous alluvial expanse in eastern England, but even so, their surface is diverse in both composition and morphology (fig. 10.2, p.119). The alluvium, together with fringing accumulations of other waterlain deposits, rests on an eroded surface of Upper Jurassic clays and older glacial drifts except to the northeast of Boston, where a depositional surface of younger glacial drifts exists. This emerges in the form of low morainic ridges at Stickney and East Keal. The peaty surface of the southern Fens around March and Ely is broken by numerous low 'islands', for example the 'Isle of Ely', some gravelly, some of till and others of bedrock, but none higher than 37 m O.D. The peats extend northward through the western Fens to the Witham valley. Eastward slightly elevated marine silts extend south from Wainfleet to Boston, Spalding and Wisbech. Seaward of the silts, reclamation of salt marshes since Norman times has displaced the coastline outwards, especially between the Welland and Ouse outfalls.

The gravel-strewn *river Cam valley* (23) opens southward from the Fens and by contrast is developed mainly on Cretaceous rocks which overstep the Kimmeridge clay west of Cambridge to rest on Oxford Clay. The Chalk Rock (Upper chalk) caps a scarp slope on the south-east side that rises near Newmarket and Balsham to over 90 m O.D. and east of Royston to about 155 m O.D. The Cam rises in Essex and flows anomalously against the dip north through the chalk scarp near Saffron Walden (Sparks and West 1965) and so into the vale where it gathers the Rhee. The latter drains the almost drift-free Ashwell part of the vale eroded along the Chalk Marl (Lower Chalk) and the Gault outcrops.

The coastal lowlands

South from Flamborough Head to the Wash, the chalk Wolds are separated from the sea by clay lowlands, but in East Anglia, complex accumulations of clays, silts, sandy and bouldery gravels overlap the seaward flanks of the till plateaux of central Norfolk and Suffolk (fig. 1.2).

Within the curve of the Yorkshire Wolds lies *Holderness* (24) an area of glacial drift and alluvium mostly below 30 m O.D. The drift, in places over 60 m thick, rests on a gently inclined chalk platform that backs in the west against a former, now largely buried, sea-cliff. Westward of the river Hull which drains most of the area, surface relief on the drifts is minimal, but eastward morainic ridges and mounds are common, crossed by several streams which rise, curiously, close to the rapidly receding modern cliffs and flow generally west or southwest to the Hull or the lower Humber.

South of the Humber, much of the character of Holderness recurs in the *Lincolnshire Marsh* (25). An inner drift-based zone, also terminating against old sea cliffs, displays frequent low but definite morainic ridges (Straw 1957). Seaward, the glacial drifts are overlain by up to 13 m of grey and brown marine clays and silts now protected by dunes, banks and revetments. These alluvial flats comprise the outer marsh, part bordering the Humber estuary, and part forming the coastal tract south of Cleethorpes until it merges with the marine silt lands and freshwater clay marshes of the northern Fens.

The *Holt-Cromer area* (26) of north Norfolk is a small but individual strip of country. The Cromer Ridge, an agglomeration of till and glacially-transported masses of rock stands about 100 m O.D. along most of its length and constitutes the major relief element of north Norfolk. West of Upper Sheringham the ridge turns southwest, and around Holt lies a composite, south-sloping sheet of glacifluvial gravels, dissected deeply by the north-flowing Glaven. East of Kelling, a 'pediment' platform has developed where recession of the gravel sheet has been pronounced. Occasional gravel outliers such as Muckleburgh Hill survive, and the platform terminates in active sea cliffs between Sheringham and Weybourne, which pass westward into bluffs fronted by partly reclaimed saltmarshes inside shingle barriers.

In *eastern Norfolk* (27), the chalk is totally concealed by shingly Pleistocene crags which are overlain in turn by glacial sands, gravels and patchy tills well exposed in the low coastal cliffs. From the crest of the Cromer Ridge, the broad interfluves between the southeast flowing Bure and its left-bank tributaries fall gradually over some 33 km to about 12 m O.D. around Ludham. The river Thurne and the lower reaches of the Bure and Ant rivers are bordered by alluvial deposits which isolate Flegg 'island' and within which many of the famous Norfolk 'Broads' were excavated in medieval times and later flooded. Similar deposits occupy the Yare valley below Norwich.

Eastern Suffolk (28) is another area where the chalk lies at depth. Between the Stour, Orwell and Deben estuaries, and stretching northeast to Aldeburgh are a series of gravelly plateaux that descend from about 45 m O.D. south of Ipswich to about 15 m O.D. north of Southwold. A few streams, such as the middle Deben, are aligned along the northwestern edge of the gravels, but most of the streams run direct to the coast, and the estuaries particularly of the Deben and the Alde are bordered by reclaimed marshes and tidal mudflats.

Clearly, eastern England is lacking in neither physiographic diversity nor geomorphic complexity. Although strong geological influence in the repetition of cuesta and vale and on the existence of many minor features can be demonstrated, the adjustment of land form to geology is far from complete. A polygenetic landscape permits a reconstruction of its evolution, and three main periods in the physiographic evolution of eastern England are readily apparent. Postglacial (Flandrian) deposits overlie 14% of eastern England and contemporaneous fluvial, aeolian and coastal processes must be mentioned separately. Glacial features abound in the area, and the various Pleistocene glaciations are collectively responsible for much of the detail of the landscape. However, the major morphological features are preglacial and retain enough of their original character to permit a rational discussion of the late Tertiary landscape.

These several considerations now concern the rest of this section, and it is convenient to work retrospectively, though reserving discussion of Flandrian events to the end.

3 The Devensian glaciation

The Devensian glaciation, even in eastern England, was a complex affair, and the drifts were emplaced during several ice advances. None of the hill areas was glaciated completely, but at the maximum stage, Pennine ice reached a little south of the Escrick moraine in the Vale of York. Along the east coast, however, ice from northern England and eastern Scotland swept a notable 100 km further south to north Norfolk. Equally remarkable is the very low gradient of the drift-limit which presumably reflects that of the ice surface. From 37 m O.D. in north Norfolk, the drift-limit rises to 114 m O.D. in the central Lincolnshire Wolds, 137 m O.D. in the northern Yorkshire Wolds, 183 m O.D. west of Scarborough and 245 m O.D. in Eskdale, that is, with an overall gradient of only 1:750, which contrasts with 1:300 for the similar limit in the Vale of York. This low gradient and the excessive southward carry of the ice are such exceptional circumstances that they can best be explained in terms of an ice-sheet surge.

The glacial stages

In east Lincolnshire and south Humberside there is abundant evidence for two stadia within the Devensian glaciation (fig. 3.1). Distinction is based on the regional contrasts in surface morphology, in the form and location of associated glacifluvial channels, and in the disposition of terminal moraines relating to two suites of drift recorded in borehole catalogues over much of the Lincolnshire Marsh (Straw 1957, 1958, 1961b). It is the Lower Marsh drifts which reach the greatest heights (114 m) on the Wolds to the west, and extend into the Humber and Wash gaps to terminate in the morphologically simple Horkstow and Stickney moraines respectively. They are now somewhat dissected and denuded on the Wolds and are associated with degraded meltwater channels (side slopes less than 15°). The Upper Marsh Till and related outwash occur only in the central Wolds and reach to 80 m O.D. but choke many valley floors, produce morainic hills and ridges and are associated with steep-sided (20° to 25°) meltwater channels and many permanent diversions of drainage. On the marsh it builds broader morainic complexes; the Hogsthorpe moraine in the southern marsh between Aby and Ingoldmells, and the Killingholme moraine in the north between Keelby and East Halton. Each of these moraines is paralleled by contemporaneous glacifluvial gravel trains that extend meltwater channel systems within the Wolds (Straw 1961b). In the south, the gravels lead southeast from Aby enclosing the leading morainic hummocks. West of the Killingholme moraine for some 13 km the

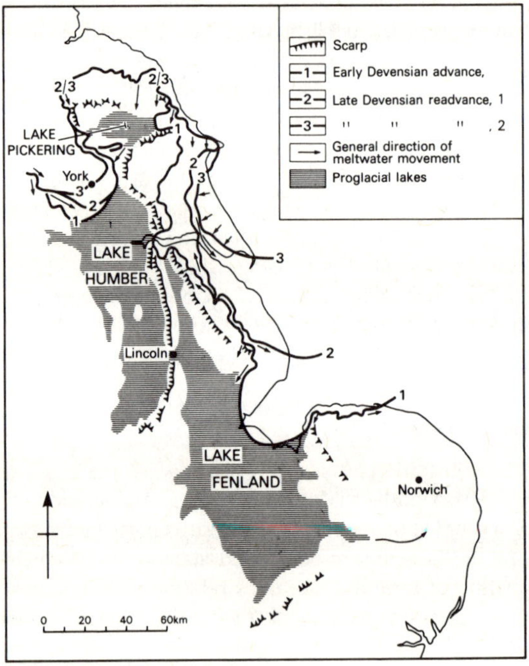

Fig. 3.1 Postulated limits of Devensian ice advances into eastern England, and maximum extent of pro-glacial lakes.

gravels descend northward in a shallow flat-floored valley toward the Humber estuary. The exceptional character of these gravels, and of related sediments within the Wold valleys, derives from their molluscan fauna which is so dominated by marine species that they were designated 'marine gravels' in the Old Series Geological Survey Memoirs.

In north Norfolk there is but one suite of drift associated with meltwater channels, weak morainic mounds and in Hunstanton Park, a sinuous esker some 2 km long (Straw 1960). The drift lies in the floors of valleys cut into chalk through a sheet of older glacial till and is banked against a former sea cliff (fig. 3.2). Southwest of Hunstanton, the drift passes into the Heacham moraine, which

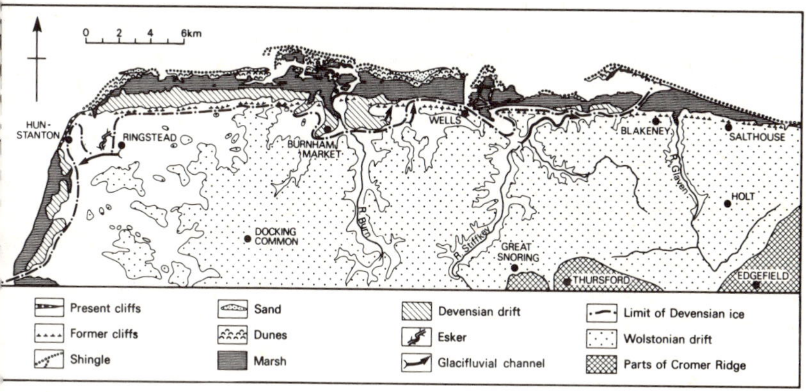

Fig. 3.2 Glacial features in north Norfolk.

matches closely the Stickney moraine across the Wash and is directly correlative with the Lower Marsh drifts in east Lincolnshire. The morphologically simple moraines of the Stickney stage together with the Hunstanton esker indicate stagnation of at least the marginal zone of the Devensian ice soon after reaching its maximum extent, which incidentally supports the ice-surge supposition. After this early advance, the North Sea ice seems to have receded for a distance and time that cannot yet be determined precisely. The Hogsthorpe-Killingholme moraines confirm unequivocally an important though less extensive readvance (fig. 3.1). The greater complexity of these moraines betrays an active ice front and different glaciological conditions.

In Holderness, Devensian drifts comprise the Drab and Purple Tills, the weathered surface layers of which were formerly designated the Hessle Till (Bisat 1939a; Catt and Penny 1966; Madgett 1975). They rest on an older Basement series, which was wave-planed during a marine transgression in the Ipswichian interglacial and perhaps also during a Mid-Devensian interstadial, and which now survives seaward of the old Sewerby-Hessle cliff. Penny (1964) identified no interstadial within the coastal sections of Devensian drift and interpreted the tills as the melt from a sandwich of ice-sheets. In places, however, the Purple Till rests in hollows in the Drab, possibly erosional (Bisat 1939b) and sands and silts intervene elsewhere. Inland there are undeniable areal contrasts in surface morphology. West of the Hull alluvium, the drift is a thin formless sheet, pared back in the valleys but spreading up to 75 m O.D. along Wold interfluves toward glacifluvial channels of rather wide and shallow form, as south of Middleton-on-the-Wold (de Boer 1945). This subdued character, reminiscent of the Lower Marsh drifts in Lincolnshire is maintained north to Driffield.

East of the river Hull the glacial surface is an assemblage of morainic and gravelly ridges and hummocks with enclosed hollows, traversed by several valleys terminating southwestward in fans of gravel and sand. At Kelsey Hill and Keyingham, the gravels contain a large derived fauna including marine and freshwater molluscs, and terrestrial and marine vertebrates. Significantly the molluscan assemblage resembles closely that of the 'marine gravels' in east Lincolnshire. All these features have been described by Valentin (1957) who traced a westward limit from Burton Agnes south to Brandesburton, Bilton and Patrington and eastward into the North Sea. This line encloses the known extent of the Purple Till (Madgett 1975) and crosses the coast at Dimlington where a thickened mass of drifts involving disturbed Basement Series occurs. This limit is believed to determine a second readvance of the Devensian ice, possibly when the Purple Till was deposited and when meltwaters coursing to the southeast finally established the modern alignment of the river Hull and lower Humber. The limit of the Lincolnshire readvance stage is less easy to recognize in Holderness, but morainic ridges around Routh, Wawne and Sutton, more definite than any features to the west, even though partly surrounded by alluvium, may relate to this stage (fig. 3.1).

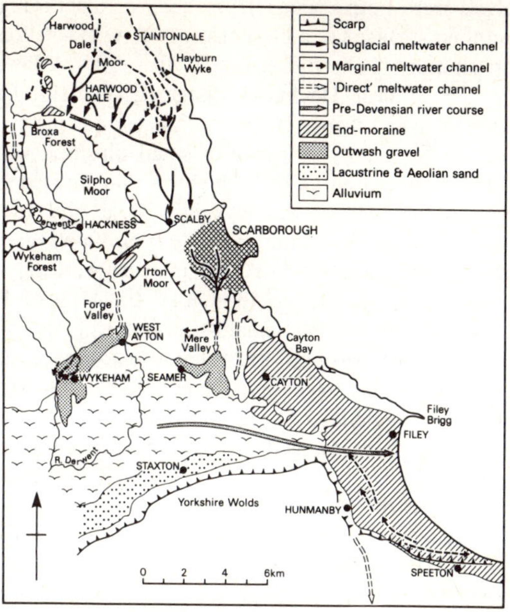

Fig. 3.3 Devensian glacial features in the Scarborough area.

North of Flamborough Head only two glacial stages can be identified on geomorphological criteria (fig. 3.3). West of Filey a fresh morainic complex, partly buried by alluvium, passes north of Cayton into thick drifts in Scarborough and Scalby. South of Filey, the morainic ridges are so compressed into the curve of the chalk scarp between Hunmanby and Speeton, where they have been mistaken for drumlins (Penny and Rawson 1969), that the most southerly ridges rise on to its crest to produce the Speeton moraine. Valentin (1957) related this moraine to the fresh features of east Holderness, so that at Flamborough Head, and to the north of it, the second readvance appears to have overlapped the first (fig. 3.1). West of Cayton the drift is predominantly of sand and gravel and constitutes a large fan, prograded east and south of Seamer by meltwaters radiating from the Mere valley, a smaller spread south of East Ayton built by Forge Valley waters and a prominent kame terrace and fan between West Ayton and Wykeham Park. These latter gravels were both eroded from and carried through Forge Valley and their surface lies a little below 75 m O.D. at West Ayton grading to an early 'Lake Pickering' at 70 m O.D. A lower gravel terrace, preserved near Hackness village, grades through Forge Valley to the East Ayton gravels. On this and supporting evidence, it seems that Forge Valley, originally a small dipslope valley, was enlarged into a superb gorge by meltwaters that utilized it on two separate occasions, first at the Wykeham stage and later at the Cayton-Speeton stage. The later meltwaters probably graded to a lower Lake Pickering level (*c* 45 m O.D.) in which were deposited the lacustrine sands that on the south side of the Vale of Pickering westward of Flixton overlie till laid down during the Wykeham advance. In the Scarborough area, therefore, there is an earlier Wykeham stage relating to the maximum advance into Holderness and Lincolnshire, and a later Cayton-Speeton stage which is considered to represent a genuine readvance and not merely a temporary stabilization of the ice front during recession (Penny and Rawson 1969). Further north, the final diversion of Jugger Howe Beck through Langdale is probably related to the Cayton-Speeton readvance state (fig. 3.3), for the fresher drift protrudes to the head of Harwood Dale. It is likely therefore that integration of the Derwent river system above Ayton was the cumulative result of glacifluvial activity in both glacial stages, rather than during one (Kendall 1902).

Northward, around Robin Hood's Bay and Whitby, in Eskdale and in Teesdale the glacial landforms seem less complex, in that they relate solely to the readvance stage. This appears to have overlapped the older stage at Scarborough, a circumstance supported by Penny's (1974) correlation of the Durham and Teesside tills with those of east Holderness. Gregory (1965) re-evaluated much of Kendall's work in Eskdale, and demonstrated decay of ice, following stagnation, in many parts during deglaciation, but there is no denying the efficacy of meltwaters in breaching the head of and modifying Newtondale before debouching into the Vale of Pickering to build the gravel delta on which Pickering stands. West of Whitby many of the gravel tracts, as on Roxby Moor, east of Liverton, and around both Guisborough and Stokesley are also a consequence of down-wasting of the ice. The till area between the Tees and the Swale carries a fresh hummocky surface, with some of the mounds being drumlinoid and the minor drainage displaying a distinct north-south trend. However, from Darlington to Middlesborough and beyond, because laminated clays were deposited on the till during a late stage in deglaciation of the Tees valley (Agar 1954) the drift surface is remarkably smooth.

In 1956 Best examined the proglacial drainage along the northern and western parts of the Cleveland Hills, and concluded that glacifluvial channels were excavated during two glacial episodes, the younger channels being associated with an ice advance during which reddish tills were emplaced below 125 m O.D. as far south as the York and Escrick moraines.

In the southern Vale of York three possible Devensian stages are indicated by the York and Escrick moraines and the Linton-Stutton gravels (fig. 3.1) and a link with the coastal stages can be achieved by reference to glacial Lake Pickering. East of York and Escrick, the simple curving moraine ridges converge at Claxton, but separate again northwest of Flaxton and run parallel about 1.6 km apart to Oulston (fig. 3.4). The Escrick moraine rises to about 90 m O.D. near Brandsby and outside it the ice-guided Bulmer Beck flows southeast to join the Derwent at Howsham. Replaced briefly by gravels around Oulston, the till ridge recurs in the Coxwold gap as

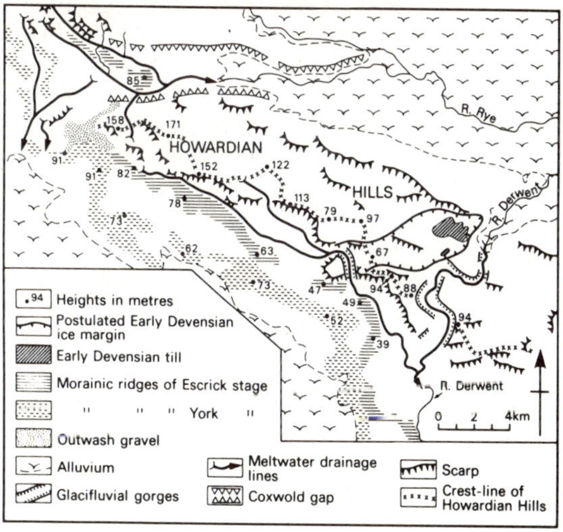

Fig. 3.4 Devensian glacial features around the eastern side of the Vale of York.

part of the Ampleforth moraine, flanked again by marginal drainage. The York moraine passes to Crayke and Husthwaite but keeps west of Coxwold. Generally these moraines have been taken to indicate the positions of the ice when the Kirkham gorge of the Derwent was formed, but this is difficult to accept. The gorge is not incised within the lowest available part of the Howardian Hills west of Welburn, but transects a high ridge reaching 75-80 m O.D. at Kirkham, to the southeast. A barrier prohibiting overflow of Lake Pickering waters westward of Welburn must be postulated. Patches of till near High Hutton confirm the presence of ice on the Howardians at a stage earlier than the construction of the Escrick moraine. This early stage could equate with Best's high-level stage, and, because the 70 m O.D. level of Lake Pickering must have existed at this time, with the Wykeham stage at the east end of the Vale.

West of the river Ouse, the York and Escrick moraines are convergent and more complex (fig. 3.5). The spectacular southeastward diversion of the river Wharfe at Wetherby belongs to the Escrick stage. Southwest of the Wharfe and reaching to almost 65 m O.D. on the Magnesian Limestone are discontinuous spreads of till and gravel which Edwards (1938, 1950) referred to as evidence of an Early Main Dales glaciation. Of these, the Linton-Stutton gravels, composed mainly of Carboniferous rocks, form a discontinuous ridge which commences 2 km west of Wetherby and runs southeast by Linton through Bramham to Stutton south of Tadcaster. Edwards *et al.* (1950) interpreted the ridge as a kame belt but only the southernmost mounds between Bramham and Stutton, built of disturbed but bedded deposits up to 8 m thick and ranging from silts to very large bounders, have a convincing kame form with steeper northern and northeastern slopes. Northwest from Bramham the gravel ridge is narrow but straight for 7 km and is a submarginal esker developed beneath a Nidderdale glacier which was deflected southeast over Harrogate to Bramham Moor on the west side of Vale of York ice during a pre-Escrick Devensian stage. The southern limit of ice in the Humberhead area at this earliest stage is not known, though Gaunt (1976) and Palmer (1966) claim extensions as far south as the Isle of Axholme and Doncaster-Gainsborough respectively. A more conservative limit is preferred here (fig. 3.1), recognizing the dearth of Devensian tills to the south of the Triassic hills of Hambleton Hough, Brayton Barff and at Holme-on-Spalding Moor.

From this review of evidence for Devensian ice limits has emerged

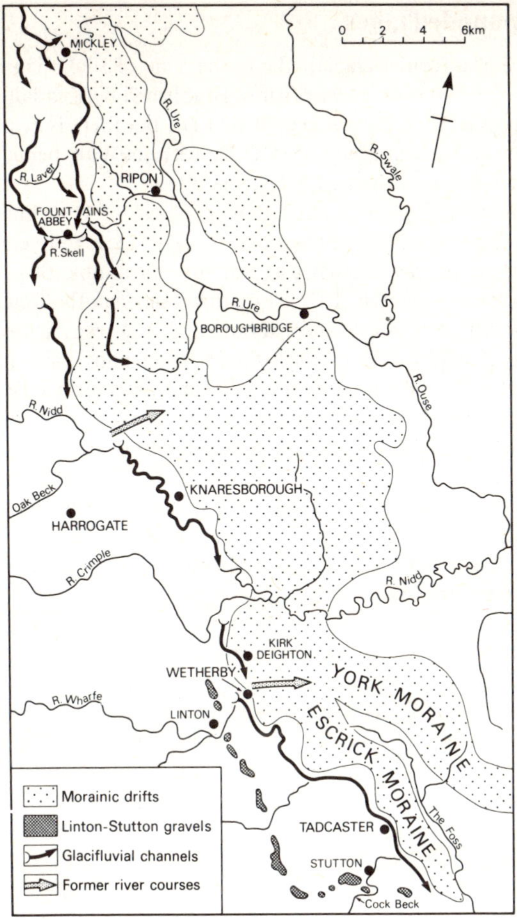

Fig. 3.5 Devensian glacial features around the western side of the Vale of York.

the proposition that in eastern England glaciation was multiple, consisting of an early advance to a maximum limit and, following a period of recession, at least two less extensive readvances. Before discussing the timing of these events within the Devensian, a brief consideration of certain extra-glacial features is necessary.

Ice-impounded lakes

Existence of a former glacial lake in the Vale of Pickering dammed by inland and coastal ice was firmly established by Kendall in 1902. The highest water level reached 70 m O.D., but there is evidence for a later lower stand at about 45 m O.D. with the Vale being emptied completely between the two lacustrine phases.

The histories of glacial Lake Humber (Lewis 1894; Melmore 1935) and its complement Lake Fenland have never been fully elucidated, though their former existence is not now in doubt (Straw 1963a; Palmer 1966; Gault *et al.* 1972). Equalized through the Lincoln gap, the maximum level of the lakes, about 30 m O.D., was achieved only when Devensian ice closed both the Humber and Wash gaps, and was apparently controlled by an outlet across central East Anglia, where lake overflow could course down the Waveney valley (fig. 3.1). Strandline features near the maximum level are mainly depositional, indicating the ephemeral and low energy characteristics of the lakes, though Harrod (1972) records a narrow bench at 25-32 m O.D. cut into slopes rising from the western Fens. Most of the deposits shown on fig. 3.6 accumulated near ice-fronts or where rivers entered the lakes.

Of particular interest are the partly dissected kame-terrace along the southern margin of the Lincolnshire Wolds near the Stickney moraine, and delta-head sands 12 km to the west on Kirkby Moor, both near 30 m O.D. The terrace was built by waters draining along the ice-edge from a meltwater system within the southeastern Wolds (Straw 1957). West of the river Bain cross-bedded sands with thin flint gravels on Kirkby Moor relate lithologically and geomorphologically to the prominent Hemingby Terrace within the Bain valley above Horncastle (fig. 3.7).

Within the Humberhead area, deposits worthy of note are those at Mill Hill, Elloughton, where gravels overlie sands containing mammalian fossils (de Boer, Neale and Penny 1957) and ridge-crest gravels at Towton at 30 m O.D. southwest of Tadcaster, which accumulated as a deltaic continuation of the kame at Wingate Hill (Palmer 1966), part of the Linton-Stutton chain. These latter gravels are the northernmost of a series of strandline sediments (Edwards 1937) that extend southward into the Doncaster area where Gaunt *et al.* (1972) designated them the Older Littoral Sands and Gravels. Between Newark and Lincoln, Beeston Terrace gravels were probably laid down by the

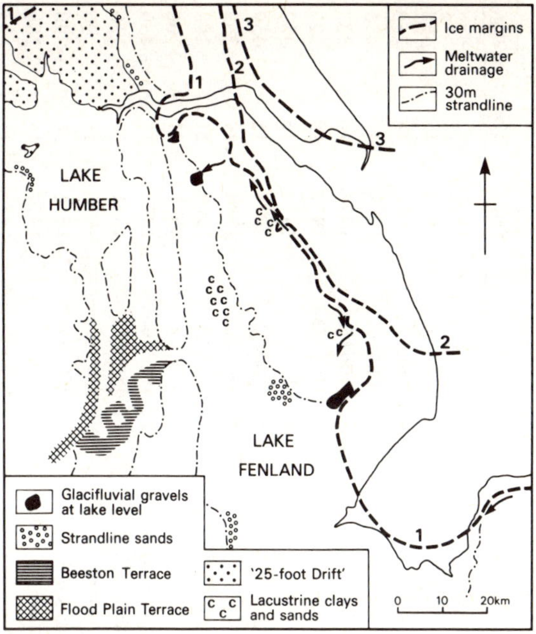

Fig. 3.6 Devensian lacustrine features in and around the Humberhead, lower Trent and Fen areas.

Trent during the lake's post-maximum phase (Straw 1963a).

The presence of low-level laminated sands, silts and clays in the Humberhead area provides a major contrast with the Fen basin. This '25-foot Drift' was deposited during a later and lower lacustrine stage with a water level of about 15 m O.D. hereafter referred to as Lake Humber II (table 3.2). It is clear that a lake did not exist in the Fen basin at this time, and therefore the Wash gap was not then closed by ice.

Lacustrine strandlines at about 30 and 15 m O.D. can be explained either as different levels of one body of water or as evidence for impounding of water on two separate occasions, in either case the higher level being the older. The two levels must relate to the two glacial stages recognized in east Lincolnshire, the higher when ice reached Norfolk, and the lower when ice stood at the Killingholme-Hogsthorpe moraine. An important corollary is that the fossiliferous

Table 3.1 Radio-carbon dates

^{14}C *years B.P.*	*Reference number*	*Locality*	*Map Reference*	*'Radio-carbon' entry*		
				Date	*Vol.*	*Page*
9,950 ± 180	N-820	East Moor	SE 607640	1972	14	227
10,280 ± 120	BIRM-349	Messingham	SE 913038	1973	15	458
10,550 ± 250	BIRM-707	Messingham	SE 916047	1976	18	258
10,700 ± 190	N-488	East Moor	SE 608640	1969	11	455
11,000 ± 200	N-963	East Moor	SE 610641	1974	16	332
11,100 ± 200	N-810	West Moor	SE 64840718	1972	14	227
11,200 ± 160	N-964	East Moor	SE 610641	1974	16	332
11,260 ± 100	BIRM-345	Grimston	TA 289352	1973	15	459
11,380 ± 260	BIRM-506	Speeton	TA 145760	1975	17	259
18,240 ± 250	BIRM-108	Dimlington cliff	TA 393217	1969	11	265
18,500 ± 400	I-3372	Dimlington cliff	TA 386224	1969	11	66
19,300 ± 300	BIRM-350	Docking Common	TF 791356	1973	15	459
19,500 ± 650	Q-590	Barnwell Station	TL 459568	1964	6	116
21,835 ± 1660	IGS-C14/96	Brantingham	SE 93852918	1974	16	96
24,000 ± 550	BIRM-412	Docking Common	TF 790357	1974	16	290
30,800 ± 360	BIRM-448	Tattershall	TF 210570	1974	16	288
34,800 ± 1000	BIRM-250	Kirkby-on-Bain	TF 532605	1973	15	3
38,600 $\pm^{1720}_{1420}$	NPL 163 B	Oxbow, Aire valley	SE 36183206	1969	11	131
39,400 ± 800	BIRM-450	Tattershall	TF 210570	1974	16	289

$42,100 \pm ^{1400}_{1100}$	BIRM-398	Tattershall	TF 210570	1974	16	288
$42,140 \pm ^{1890}_{1530}$	BIRM-88	Earith	TL 390763	1969	11	264
$42,200 \pm 1000$	BIRM-409	Tattershall	TF 210570	1974	16	287
$43,000 \pm ^{1300}_{1100}$	BIRM-341	Tattershall	TF 210570	1973	15	453
$44,300 \pm ^{1600}_{1300}$	BIRM-408	Tattershall	TF 210570	1974	16	288
$> 40,000$	IGS-C14/125	Woodston	TL 17999608	1974	16	101
$> 40,000$	IGS-C14/129	Setch, Norfolk	TF 65021419	1974	16	102
$> 40,000$	IGS-C14/73	Dunston Common	TG 22700267	1973	15	300
$> 40,000$	IGS-C14/74	Dunston Common	TD 22700267	1973	15	300
$> 42,000$	BIRM-260	Tattershall	TF 208569	1973	15	4
$> 45,000$	BIRM-251	Kirkby-on-Bain	TF 532605	1973	15	3
$> 45,000$	BIRM-86	Earith	TL 390763	1969	11	264
$> 47,600$	BIRM-365	Hoxne	TM 177766	1973	15	460

gravels west of the Killingholme moraine and at Kelsey Hill and Keyingham are broadly contemporaneous with Lake Humber II and correlate stratigraphically with the '25-foot Drift' (table 3.2).

The sequence in the Fen basin

At present, the Devensian cold stage in Britain is resolved into three phases defined by radiocarbon dates (Mitchell *et al.* 1973). The maximum advance of ice in northern England is usually placed, on the basis of northwest Midland evidence, within the Late Devensian (Sparks and West 1972; Penny 1974). In eastern England support for this view rests on two assumptions; that the Holderness coastal tills, above mossy silts radiocarbon-dated at about 18,250 B.P. (table 3.1), represent the full Devensian sequence, and that the youngest ice advance into Holderness was also the one to reach Norfolk. It has been shown however (Straw 1961b, 1969) that the east Holderness tills correlate only with the Killingholme-Hogsthorpe moraines, and Suggate and West's (1959) report of Late-Glacial kettle-hole sediments on the Hogsthorpe moraine near Aby confirm a Late Devensian date for this feature. The maximum advance in east Lincolnshire and Norfolk is a considerably older event, for the Lower Marsh tills on the Wolds are weathered thoroughly and dissected, and are associated with meltwater channels of degraded form (Straw 1961b).

The relative age of this event can be established in the lower Bain valley area, where low-lying fluviatile gravels around Tattershall have yielded organic remains from clayey and silty lenses radiocarbon dated between 30 800 and 44 300 years B.P. (table 3.1). The gravels, mainly Middle Devensian, contain many intra-formational periglacial structures, and cold climatic conditions for most of the period are confirmed by the organic material, though at least one short interval of very warm climate is evident (Girling 1974). The gravels rest on pre-Devensian till, except where thin Ipswichian sediments intervene (Girling 1974) and can be traced up the Bain valley for some 20 km in low terraces covered in part only by Flandrian alluvium. South of Tattershall where their surface lies at about 5 m O.D., they protrude fanlike beneath the Fen peat. North of Tattershall, the Bain valley is contained within low bluffs of chalky till above which on the west, lie the deltaic Kirkby Moor sands (fig. 3.7). The latter and the correlative Hemingby Terrace in the Bain valley are clearly older than the Tattershall gravels which contain no evidence for Lake Fenland and an Early Devensian date for the lake and therefore for the ice advance to the Stickney moraine is firmly indicated (table 6.1).

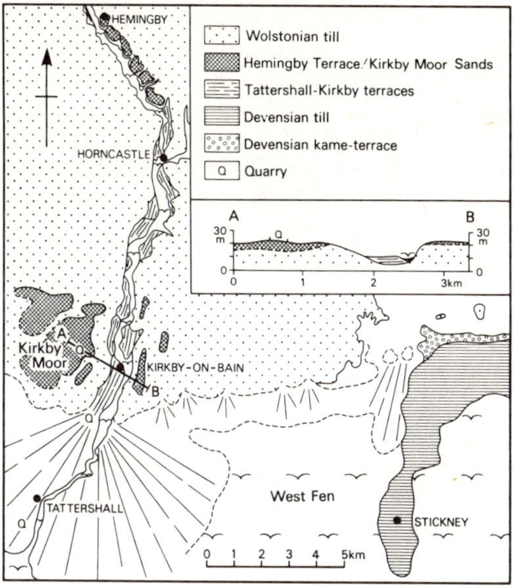

Fig. 3.7 Early and Middle Devensian terrace features in the lower Bain valley around Tattershall.

The Tattershall deposits, which represent aggradation over at least 15 000 years, are not unique in the Fen basin. Similar fans passing upstream into low terraces flair out of the Welland, Nene and Ouse valleys in the western Fens, and from the Cam, Wissey and Nar valleys on the eastern side (fig. 10.2, p.119). Gravels in the lowest Ouse terrace at Earith have been described by Bell (1970). Depositional structures are closely comparable with Tattershall, and a radiocarbon date of 42 140 years B.P. (table 3.1) has been obtained for a contained silt later. Together the Intermediate (Second) and First terraces of the Cam span a period between late Early Devensian and early Late Devensian (Lambert, Pearson and Sparks 1963; Coope 1968), and Early Devensian terrace gravels of the river Wissey survive at Wretton in southwest Norfolk (Sparks and West 1970; West *et al.* 1974). Prentice (1950) discussed briefly the distribution of Fen Margin gravels in the western Fens, noting the Welland gravel fan, and correlated them generally with the Nene and Bain gravels.

For present purposes the Wretton gravels may be compared briefly with the Tattershall gravels, deposition of which it has been argued took place after the occurrence of Lake Fenland. At Wretton, some 6 m of Early Devensian gravels rest unconformably on Gault Clay and relics of Ipswichian sediments. Layers of organic muds and well-sorted sands are interbedded with the gravels, and certain horizons display periglacial involutions, ice-wedge casts and ground-ice structures. However, the uppermost metre of sands and gravels is unfossiliferous and extends over the whole terrace. It is suggested here that these upper sediments relate to Lake Fenland, submergence by which suppressed further fluvial sedimentation at Wretton, for on draining of the lake the Wissey incised a new course to the south. A very cold climatic phase toward the top of the sequence is not supported by pollen spectra, but fossil *Coleoptera* (Coope, Appendix 1 in West *et al.* 1974) do indicate a relatively barren tundra with a more extreme cold continental climate prior to deposition of the capping materials.

It appears therefore that as sea-level fell from the Ipswichian 'high' and as Devensian tundra environments became established so vast quantities of rock waste from the Fen river catchments were distributed widely over the floor of the Fen basin, interrupted only in late Early Devensian time by a relatively short-lived lacustine submergence to 30 m O.D.

The persistently low gradients of deposition displayed by the various fans may in part be a response to relatively high base-levels associated not only with the lacustrine phase but also with sea-level during the Middle Devensian. Evans (1971) suggests that a sea-level as high as 2-3 m O.D. may have occurred at this time, and Donovan (1962) argued from Bristol Channel evidence that sea-level oscillated around its present level between 45000 and 24000 years ago, but Thom (1973) after a thorough discussion of interstadial sea-levels suggests a level of —10 m (± 10 m) O.D. A marine submergence of areas previously overspread by Early Devensian till eastward of Lincolnshire provides a satisfactory explanation for the incorporation of marine molluscs in the outwash gravels associated with the Late Devensian Hogsthorpe moraine.

The sequence in the Humberhead area

Within this area, Devensian drifts are ascribed almost entirely to the Late Devensian (Mitchell *et al.* 1973; Penny 1974). Palmer (1966) claimed Devensian ice to have reached to Doncaster and Gainsborough and to have retreated to the York moraine in Late-Glacial time (late Late Devensian).

At Austerfield near Doncaster, Gaunt *et al.* (1972) identify Older River Sand and Gravel as Ipswichian, separated by a 'lower periglacial surface' from Older Littoral Sand and Gravel. The latter, and the '25-foot Drift', parted by an 'upper periglacial surface', relate to Lake Humber. Further north at Langham and Westfield, Ipswichian sediments correlated with the Older River Sand and Gravel are succeeded by a horizon representing the lower periglacial surface and at Langham by up to 4.6 m of upper gravel not related to Lake Humber (Gaunt *et al.* 1974). The geomorphological relationships of these deposits in the country around Austerfield, Snaith and Armthorpe testify to the phases of valley deepening between Wolstonian drifts and the Ipswichian sediments, and between the latter and the '25-foot Drift'.

At the Oxbow opencast site, a series of mainly fluvial sediments lies within an Aire valley deepened by post-Wolstonian erosion (Gaunt *et al.* 1971). At the base, a Lower Silt overlying 1.5 m of Lower Sand and Gravel contained a mammoth tusk dated at 38 600 radiocarbon years B.P., placing it firmly within the Middle Devensian. Succeeding Middle Sand and Gravel carried a Middle Silt, and an upper group of deposits was shown to be of Flandrian age. Glacifluvially transported material in the Middle Sand and Gravel was considered to support a correlation with a presumed Late Devensian glaciation in the upper Aire valley.

Gaunt (1976) has given reason for an ice advance to the Isle of Axholme associated with Lake Humber I, and has reported a radiocarbon date of about 22 000 years B.P. on bone discovered in what are purported to be lacustrine sands and gravels at 25 m O.D. near Brantingham, northwest of the Humber gap.

This collective view denies the existence of Middle or Early Devensian sediments in the Humberhead area except for the Aire valley Lower Sand and Gravel and Silt, and creates a major chronostratigraphic hiatus between the Ipswichian and the Late Devensian deposits. Thus some 40 000 years are 'lost', and all recognized

Devensian events are placed within about 10000 years of Late Devensian time — an extraordinary situation by comparison with the Fen basin.

Though Gaunt *et al.* (1972) incline to an Ipswichian age for the whole of the Older River Sand and Gravel, it is clear that much of it above the Ipswichian silts could be, as at Wretton, of Early Devensian age (table 6.1). Geomorphologically, the fan-like spread of these gravels and sands from the Idle and Don valleys are identical to those around the Fen basin. The Lower Sand and Gravel in the Aire valley may also be Early Devensian, and there is no apparent reason why these deposits should not constitute both the upper and lower terraces of the Aire which were partly a response to the rising level of Lake Humber I. Incision into these gravels, controlled by lowering of the lake, was followed through the Middle and Late Devensian by deposition of the Lower Silt, Middle Sand and Gravel, and Middle Silt which, at 14-16 m O.D., accords closely with the level of Lake Humber II. The upper gravel at Langham would relate conveniently to the Middle Sand and Gravel.

These considerations suffice to demonstrate that Early and Middle Devensian fluviatile and lacustrine sediments do exist within the Humberhead area in locations and on a scale comparable with the Fen basin, and further that Lake Humber I was more likely a late Early Devensian, rather than a Late Devensian event (table 6.1). The antiquity of this lake stage is confirmed by the morphology of features pertaining to it. Whereas the '25-foot Drift' still presents vast undissected surfaces, is overlain only by Flandrian alluvium and passes up valleys into very low terraces, the littoral deposits of Lake Humber I between Doncaster and Tadcaster all occupy interfluve positions parted by valleys draining off the Magnesian Limestone. On the east side Lake Humber I gravels lie in elevated positions on benches or hill tops. In the Fen basin, the Kirkby Moor sands and the related Hemingby Terrace have been similarly converted into interfluve or spur-tip positions. Geomorphologically therefore it is unreasonable to compress both stages of Lake Humber into a very few thousand years.

The Trent sequence supports this interpretation. In late-Wolstonian time the Trent excavated a new course from Long Eaton to Newark and northward to Gainsborough (Straw 1963a) and followed this course through the Ipswichian. Its temporary diversion to Lincoln as witnessed by the lower section of the Beeston Terrace relates to Lake Humber I. During the Middle Devensian it commenced aggradation of the Flood Plain Terrace, a process prolonged by the rising waters of Lake Humber II until overflow was possible over the Beeston Terrace gravels through the Lincoln gap. Incision following reopening of the Humber gap eventually restored the Trent to its present course.

The argument that Lake Humber I and the ice advance which closed the Humber and Wash gaps simultaneously were Early Devensian phenomena is further supported by the glacial sequence in eastern Humberside, and here the fossiliferous gravels which occur only east and southeast of the Humber gap once again assume significance. The Kelsey Hill, Keyingham, and Brocklesby-Goxhill outwash gravels relate to the Late Devensian advance to the Killingholme moraine. They contain a derived but mixed fauna of Ipswichian and Devensian forms, best explained by ice advancing over a Middle Devensian sea-floor and Humber estuary. The fossils occur in greatest profusion only east of the Humber gap (with significant quantities of *Corbicula fluminalis*), but were also carried southward by the ice into east Lincolnshire. The gravels rest on that till which reaches to the Horkstow moraine, arguably of Early Devensian date.

This convergence of evidence from the Fen basin, the Humberhead area and the east coast is sufficient to postulate with confidence that the Devensian cold stage is fully represented in eastern England (table 6.1), as maintained previously by Edwards *et al.* (1950), and that in apparent contrast with Wales and the north and west of England, the Early Devensian advance was, south of Scarborough and in the Vale of York, more extensive than that of the Late Devensian.

Periglaciation

Devensian features of periglacial character are widespread in eastern England (fig. 3.8) and merit a much longer discussion than is possible here. Subsurface structures and cryoturbation deposits are common particularly beyond the Late Devensian glacial limit. Structures include ice-wedge casts, sand wedges, involutions and vertical erection of stones with disturbance of bedding in both bedrock and drifts, gulls, step-faulting and folding related to cambering, and faults, micro-folds and displacement of deposits resulting from the growth and decay of lenses and layers of ground ice.

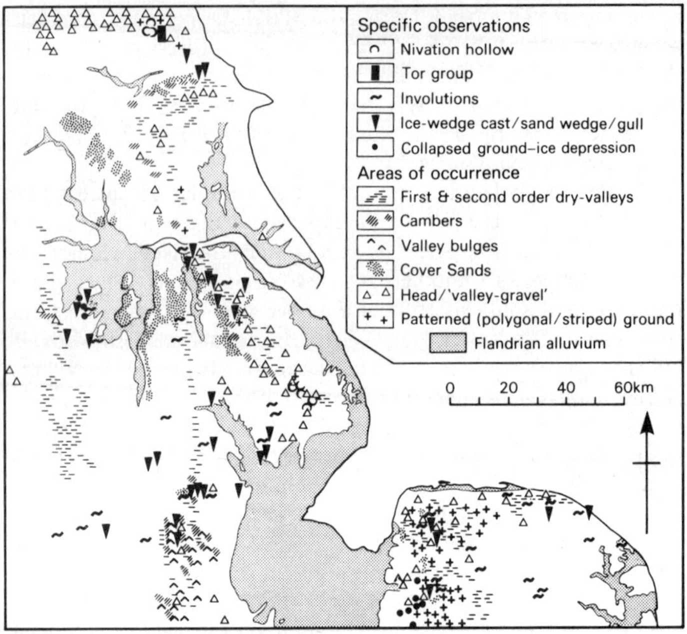

Fig. 3.8 Provisional map of Devensian and early Flandrian periglacial landforms and near-surface deposits and structures. (Based on personal field work and references cited.)

The smaller structures rarely have physiographic expression, though they have been detected frequently both in the field and on aerial photographs by their influence on present-day vegetation and soils. Stripes and polygonal patterns have been shown to cover large areas of western Norfolk and the Brecklands (Watt 1955; Watt, Perrin and West 1966; Williams 1964, 1968, 1969) and have been observed also in the North York Moors (Dimbleby 1952). Exposures in Devensian glacifluvial and fluvial sands and gravels display evidence of surface and intra-formational cryoturbation in many horizons, and reveal casts of ice-wedge roots which are particularly diagnostic of former permafrost conditions. On the gentle slopes of platform and bench sites on limestone and sandstone rocks, cryoturbated surface layers up to 1.5 m depth are also common, but on slopes of more than a few degrees, denudation by solifluction has carried frost-riven waste to lower levels.

Larger structures, however, associated with cambering and ground ice formation do have morphological expression in some localities. Excellent examples occur in the upper Witham and other valleys in the limestone areas of southwest Lincolnshire (fig. 3.8). Breakaway joint-bounded blocks and slabs at the trailing edge of camber flaps give rise to shallow slump and slide features as along parts of the scarps of the northern and northwestern Yorkshire Wolds, the central Lincolnshire Wolds, and Lincoln Cliff. Valley-bulging, structurally associated with cambering, has been reported widely from southwest Lincolnshire and has influenced the development of valley meanders in certain catchments of that area (Harrod 1972). Not all such features are of Devensian age, and valley and scarp-foot bulging and cambering can develop under non-periglacial conditions, for they result essentially from the relief of loading stresses and can therefore develop as a normal consequence of the lengthening and steepening of slopes on incompetent rocks and where groundwater levels and pressures exceed critical limits. However, in eastern England, periods of valley deepening have clearly coincided mainly with cold climate conditions when sparse vegetation cover, a dominance of mechanical weathering and seasonally high-energy fluvial regimes were prevalent.

In the Thetford area (Worssam and Taylor 1969) and parts of west Norfolk (Sparks *et al.* 1972) assemblages of hollows of irregular shape and size surrounded wholly or partly by ridges up to 3 m in amplitude produce a distinctive micro-relief. The hollows, some containing peat, vary from 10-120 m in diameter, and occur over the outcrops of the Middle and Lower Chalk but are specifically developed, as at East Walton Common in Norfolk, in superficial deposits of sandy chalk rubble. Pollen analysis suggests a Late Devensian age, and their mode of formation relates to the growth and decay of ground ice, but whatever the precise mechanism an ill-drained situation and a permafrost condition would appear prerequisite. At the Fen margin they disappear beneath Flandrian peat and in the Brecklands a relationship with 'soil' stripes can be demonstrated.

East of Doncaster the Flandrian peats of West Moor are almost enclosed within a near-circular hollow, some 2.5 kilometres across, bounded by Older River Sand and Gravel (Gaunt *et al.* 1972) which are arguably of Early or even Middle Devensian age. A 14C date of 11 100 B.P. (table 3.1) on a soil beneath thin peat and blown sand together with the presence of '25-foot Drift' within the hollow confirms its Devensian age. Two similar though smaller and less regular hollows east of Barnby Dun, one north of Cantley and two more west and southwest of Branton are all developed in Bunter Sandstone, within the tract of Older River Sand and Gravel, and contain either solifluction deposits or Flandrian peat. Fluvial and glacial processes do not account satisfactorily for these features, and it is suggested here that, their large size notwithstanding, they too are collapsed ground ice forms. Noting their ideal location on highly pervious Bunter Sandstone and fluviatile gravels close to the former margins of Lake Humber, they probably developed during melting of large ice lenses or massive ice bodies that had built up with the freezing of groundwater under artesian pressure.

The mass movement of rock waste on slopes greater than 2-3° by processes of congelifraction and niveo-solifluction was widespread during Devensian times and contributed substantially to the removal of older glacial drifts from large areas. The floors of trunk and major lateral valleys of drainage systems within the limestone cuestas commonly rest on thick valley-gravels consisting of angular and sub-angular fragments of local rocks in sandy or loamy matrices, and such gravels often pass downstream into better-sorted fluviatile sands and gravels. At South Ferriby on southern Humberside low cliffs provide good sections across a soliflual terrace built of angular chalk gravel with a sand matrix. Crudely bedded, the gravels rest on presumed Early Devensian till, display frequent evidence of cryoturbation and were derived from backing chalk slopes and small dry valleys.

Many chalk dry valleys originate on plateau areas in barely detectable broad depressions that lead into shallow linear channels of convexo-concave cross-profile, and convex thalweg. As the channels deepen and collect others so the thalweg gradient becomes constant, rectilinear elements develop on the side slopes and the basal concavity is reduced. This headward morphology is consistent less with spring-sapping than with erosion by surface runoff provided by the melt of a snow cover. Seasonal melt of snow cover, surface runoff maintained by permafrost, and congelifluction processes on side slopes summarize the periglacial conditions under which minor dry valleys reached densities sufficient to bring most of the land under slope; hence the gross reduction of remnants of preglacial 'planation' surfaces and the widespread removal of pre-Devensian drifts.

Similar dry-valley forms occur on the Keuper Marl and particularly the Bunter Sandstone outcrops, where solutional and spring-sapping hypotheses are difficult to apply. On the Jurassic and Permian cuestas, however, whilst many dry-valley systems exist, the stratigraphy and alternating lithology of the rocks render spring-sapping a more important process of valley network extension, but even here, especially where limestone and sandstone strata are not pierced, presumed periglacial dry valleys with headward forms similar to those on the Wolds are to be found.

Along the chalk scarp of the southern Lincolnshire Wolds and within the Oxcombe dip-slope valley a little westward of the maximum Devensian glacial limit, a number of short, dry, round-headed coombes have the character of nivation hollows from which fans of chalky 'head' spread out over lower ground. Similar forms do not exist along the scarp of the Yorkshire Wolds (Lewin 1969), but an interesting recurrence exists within the Tabular Hills of the North York Moors, where the fused valley-heads of Levisham Beck have produced the spectacular Hole of Horcum (fig. 3.8). Here the original spring-sapped ravines have been converted into a group of six conjoint arcuate recesses from which rock waste has converged to blanket the valley floor with at least 4 m of head. The restricted occurrence of nivation hollows in eastern England suggests that, as with cambering and valley-bulging, particular combinations of environmental and geological circumstances are essential for their development.

Although some of the periglacial deposits in eastern England are old (Alabaster and Straw 1976), the majority are Devensian, but it is not always possible to establish which stage they belong to, except perhaps in two instances.

From Docking in northwest Norfolk, a shallow dry valley runs southeast as one head of the Wensum river system. Eroded to the chalk through the older till sheet of northwestern High Norfolk, flint gravels were later aggraded in its floor. At Docking Common quarry sections in the valley floor have revealed up to 10 m of sandy angular flint gravel interspersed with seams and lenses of loessic sandy silt. Three metres from the surface and within the gravel over a 50 m width, a thin sandy layer containing carbonaceous material constitutes the topmost horizon of a complex and partly reworked soil (Straw 1979). Both the overlying and underlying gravels are strongly involuted. The carbonaceous material in the topmost soil horizon has yielded radiocarbon dates of about 19300 and 24000 years B.P. (table 3.1), which indicate that the upper gravels and involutions are Late Devensian, broadly contemporaneous with the Holderness and east Lincolnshire glacial readvances. The lithological and structural similarities between the lower and upper gravels and their common geomorphic location along the present valley floor suggests indeed that they are both of Devensian age.

Across much of north Lincolnshire and parts of south Humberside lies a sheet of Cover Sands (fig. 3.8). The Sands, largely aeolian and of westerly derivation, provide some micro-relief where they emerge in low mounds from beneath Trent alluvium, and as dune and 'blow-out' country on Risby and Manton Warrens along Lincoln Edge. Eastward the Sands pass beneath the Ancholme alluvium to reappear along the foot of the Cretaceous scarp as far south as Market Rasen, and they spread onto the high Wolds near Caistor (Straw 1963b). Perrin *et al.* (1974) have demonstrated widespread aeolian deposition of sand and silt over eastern England in Late Devensian time. The Cover Sands are clearly part of this larger phenomenon, and their deposition continued into the Flandrian until more humid and closely vegetated surface conditions prevailed. A radiocarbon date of 10280 years B.P. has been obtained on peat, formed under a cold climate, from the base of the Cover Sand at Messingham, south of Scunthorpe (table 3.1).

Northeast of York, Matthews (1970) reported similar aeolian sands, and further areas lie in the Humberhead region southeast of the Escrick moraine. Matthews referred the sands to the period 10700 to 9950 radiocarbon years B.P. on assays of included peat and gyttja (table 3.1). The base of the sand is involved in periglacial structures, and surface layers have been reworked after Anglo-Saxon and Medieval forest clearance. Around York and Selby, winnowing of the sand from glacifluvial deposits to the west seems likely. Perrin *et al.* (1974) suggest that much of the sand of the Brecklands in East Anglia is derived from blown sand rather than from a local sandy till, and it is interesting to note that if this is the case then a possible source lies once more to the west in the terraces of the Cam and Great Ouse.

The dramatic events of the Devensian period were clearly formative episodes in the geomorphological evolution of eastern England, incorporating multi-stage glaciation in the north and east, repetitive periglaciation over the non-glaciated areas, flooding of the major vales, and mass transport of sediment by rivers and dusts by wind under 'Arctic' regimes, and it is unrealistic to compress them into the few thousands of years of the Late Devensian. Even so, the total effect of the Devensian on the land form of eastern England was essentially one of modification of detail, and it is now appropriate to examine the pre-Devensian legacy.

4 The Ipswichian interglacial

In most areas of their occurrence Devensian tills lie on a bedrock surface and not on older drifts. They are generally low-lying, weakly dissected, and retain constructional surface forms. The earlier glacial deposits, traditionally and collectively known as the Older Drift are by contrast well-dissected, especially outside East Anglia, often deeply weathered and smooth-surfaced. The bulk of the Devensian drift lies north of the Wash whereas the Older Drift is now restricted almost entirely to areas south of the Humber. The glacial landscape of Yorkshire thus differs in many ways from that of East Anglia, and it is in Lincolnshire that the geomorphological contrasts are best displayed.

The phase of dissection between the Older Drift and the Devensian tills would seem, on the basis of biological evidence acquired through studies of fossiliferous deposits at sites throughout eastern England (fig. 4.1), to accommodate at least one interglacial period. In a physiographic account it is not appropriate to review all the floral and faunal evidence, and discussion will centre mainly on marine and fluvial activity, but the general consensus (Sparks and West 1972) is that during this Ipswichian Interglacial, eastern England experienced a fully temperate oceanic climate, and carried a luxuriant though ecologically variable vegetation cover, supporting a richly diverse fauna. The interglacial progressed through reasonably distinct vegetation stages that conform with the model proposed by Turner and West (1968).

Several deposits of known Ipswichian age survive within the Fen basin (fig. 4.1), and West (1972) and Sparks and West (1972) have reviewed some of these in relation to possible interglacial sea-levels. Thus the marine March Gravels and the fluvial Third (Barnwell)

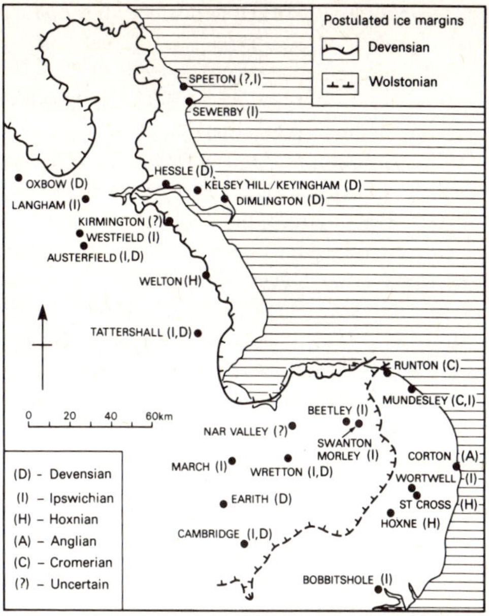

Fig. 4.1 Interglacial and interstadial sites in eastern England.

Terrace of the Cam are taken to indicate a maximum high-tide level of 12-13 m O.D., probably during the Late Temperate stage (zone Ip III). However, 45 km northnortheast of Cambridge, the Wretton site revealed brackish water deposits of Early Temperate stage (zone Ip IIb) whose present elevation is only -1.95 to +0.45 m O.D. In the western Fens, the highest terraces of the Welland, Nene, and Ouse are probably broadly contemporaneous with the Third Cam terrace, especially as some of the lower terraces are of known Middle Devensian age. On the Fen margin these terraces lie at about 15-18 m O.D., but it is not possible to establish direct connections with the March Gravels. Harrod (1972) delimited a number of weakly developed surfaces cut on Older Drift northeast and southeast of Sleaford. His Heckington Surface carries gravels at 12-18 m O.D., but the more extensive marine Fen Margin gravel lies at 3-9 m O.D. backing against steepish slopes for some 24 km south of Horbling. Harrod disagrees with Prentice (1950) by correlating these somewhat dissected and strongly frost-worked deposits with the March Gravels. At Tattershall, beneath Middle Devensian gravels, thin Ipswichian deposits rest on Older Drift at about present O.D., and a similar situation exists at Wretton where some 3-4 m of Devensian gravels overlie Ipswichian channel sediments. These facts reveal an interesting physiographic situation in that the Devensian gravels at Tattershall and Wretton *on* Ipswichian deposits, whereas in the south and southwestern Fen margins the Ipswichian fluvial and marine gravels were converted into terraces or low eminences before aggradation of the *lower*-lying Devensian gravels. To account for the slight tilting of the Ipswichian deposits relative to the Middle Devensian fluviatile gravels it does not seem unduly speculative to suggest slight relative subsidence of the Fens northeast of a line from about Sleaford to Mildenhall before the Middle Devensian and to propose a glacio-isostatic mechanism for it involving an Early Devensian ice advance into the Wash gap.

Evaluation of the Fen basin and southeast England evidence led West (1972) to the conclusion that a eustatically controlled maximum mean sea-level height in the Ipswichian was around 7-8 m O.D., and this can provide a basis for assessment of Ipswichian marine activity elsewhere in eastern England.

East of Tattershall, the southern edge of the Lincolnshire Wolds comprises a degraded marine cliff now broken by numerous valleys (Linton 1954). In part this cliff is cut across older till, but east of West Keal it is developed in Lower Cretaceous rocks. The cliff turns sharply north at Gunby (fig. 4.2) and defines abruptly the eastern margin of the Lincolnshire and Yorkshire wolds as far north as Sewerby near Bridlington where it is transected by the modern coastline. Harrod (1972) interpreted the steeper slopes behind the Fen Margin gravels as a similar cliff cut through Older Drift into Oxford Clay and Kellaways Rock and now degraded to about 3-5°. In east Lincolnshire, marine deposits are not exposed, but analysis of borehole records reveals an extensive abrasion platform beneath Devensian drifts (figs 1.2, 4.2) declining gently south and east from the cliff foot, cut across various Cretaceous strata (Straw 1961a). The back of this platform lies at about 2-4 m O.D. and in places, for example near Hawerby, slopes of chalk obscured by Devensian till up to 20° exist.

In east Yorkshire, deposits associated with the cliff have been exposed at Hessle on north Humberside, and at Sewerby. At both

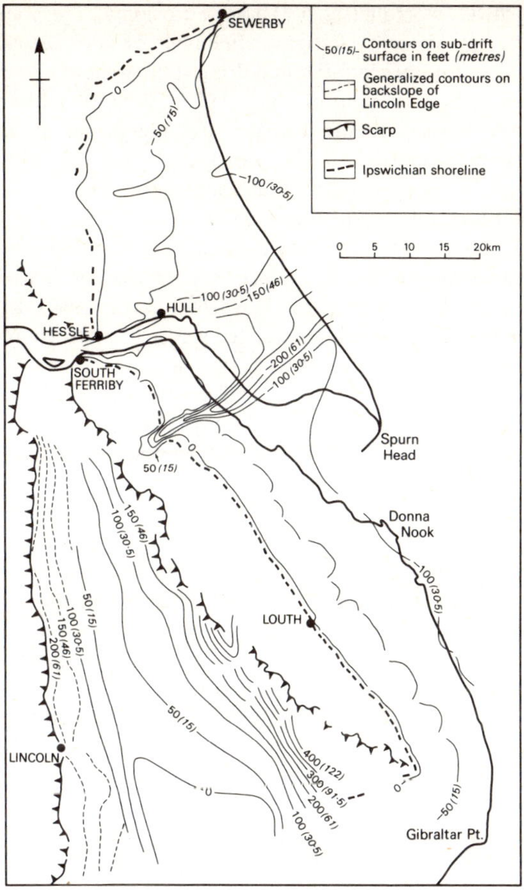

Fig. 4.2 The sub-drift surface of central and east Lincolnshire and Holderness.

sites chalky gravel and their contained faunal remains have been examined recently by Boylan (1967). He concluded firmly that the Sewerby fauna was of Ipswichian age, but at Hessle, chalky rubble and its fossils were cold climate, probably Weichselian (Devensian). However, as at Sewerby, the cliff foot lies at about 2.5 m O.D., complemented eastward by a broad abrasion platform cut across chalk and in the north the Basement Till.

In northwest Norfolk (fig. 3.2) marine erosion was responsible for producing the straight west-east alignment of the steepish chalk slopes rising from beneath the Flandrian saltmarshes and Devensian till (Straw 1960). Remnants of older glacial till on the hills and spurs behind the degraded cliff confirm its interglacial character, a circumstance supported by the Morston shingle beach (Solomon 1931) beneath the Devensian till at about 4 m O.D.

On the north side of Flamborough Head the Lower Speeton Shell Bed near present beach level is regarded as of Ipswichian age (Catt and Penny 1966) and north of Scarborough the modern coastline approximates closely that of the Ipswichian (Agar 1960). Remnants of the interglacial cliffs and the related wave-cut platform indicate a sea-level only a little higher than at present.

When these Ipswichian cliffs from Teesside to Norfolk were being cut the east coast clearly constituted a high energy environment, and within the Fen basin, cliffing took place opposite the longer fetch through the Wash gap. The Fen Margin gravels and the March Gravels confirm, however, that the Fen basin was also a repository of marine sediments, as is the Wash today.

If the mean altitude of the cliff foot from Sewerby to Morston together with the beach fragments approximates mean hightide level, and a tidal range similar to present is assumed, then a mean sea-level barely higher than present must be inferred. Comparison with the 7-8 m level deduced for the Fen basin (West 1972) would seem to confirm slight Early Devensian subsidence to the northeast. These conclusions regarding the maximum Ipswichian sea-level raise problems concerning the fossiliferous sites at Kirmington in South Humberside and in the Nar valley in northwest Norfolk.

At Kirmington (fig. 4.1) estuarine clays and peat lie beneath some 1-3 m of coarse shingle of typically chatter-marked cobbles capped in turn by up to 2 m of Devensian till. The deposits are contained within a low hill that coincides with a weak morainic ridge marking the limit of the maximum Devensian advance. The hill lies in the centre of a broad vale which reflects the headward portion of a deep infilled trough which runs northeast across the chalk platform (fig. 4.2). Boylan (1966) identified some 16 m of estuarine sediments beneath the shingle and ascribed them to the interglacial (Hoxnian) preceding the Ipswichian, but on balance palynological and palaeontological studies have been age-inconclusive. The uppermost till is undoubtedly part of the Devensian complex of east Lincolnshire and it rests directly on the marine shingle which has been strongly cryoturbated. The stratigraphical and geomorphological circumstances support an Ipswichian age for the deposits (Straw 1969). The Kirmington deposits indicate a mean sea-level approaching 22 m O.D. which is manifestly much higher than that so far adduced for the Ipswichian maximum. If they are indeed of Hoxnian age, then their survival poses a geomorphological problem, for the area was undoubtedly heavily glaciated during the Wolstonian. The deposits are a small remnant of formerly more extensive interglacial sediments that on this reckoning must have been largely removed by Wolstonian and Ipswichian erosion, but it is difficult to understand the survival of the flint shingle, when it is only overlain by Devensian till.

Southeast of King's Lynn in west Norfolk, the Nar valley draining west off the chalk expands on the fenland margin into a complex of alluvial flats, terrace features and low eminences of bedrock and glacial till. Deposits long known as the Nar Valley Beds were examined palynologically by Stevens (1959) who preferred a Hoxnian age. As most previous workers had accepted a Last (Ipswichian) interglacial date, and as freshwater deposits are overlain by an estuarine clay which indicates a maximum high-tide level to about 20 m, another controversial situation exists. The palynological evidence was not however age-conclusive, and correlation with other Hoxnian sites was achieved substantially by relating the underlying till to Lowestoft (Anglian) tills in and around the lower Waveney valley, and by identifying a one-metre thick enveloping sandy deposit as Gipping (Wolstonian) till (Stevens 1959). The latter however has every characteristic of a periglacial 'head' typical of Devensian accumulations over much of west Norfolk, and the lower till, as discussed below, is more likely of Wolstonian age, for colour and composition alone are not sufficient bases for designating the Nar Valley till as Anglian. Geomorphological objections to a Hoxnian age concern the extent and morphological relationships of the Nar Valley Beds. The latter extend over several square kilometres with a smooth terrace-like surface between low hills, and it seems inconceivable that they could retain this surface unmodified through both Wolstonian glaciation and subsequent erosion of Wolstonian drifts. Moreover, their deposition was controlled by the pattern of dissection that had affected the glacial drifts on which they partly rest.

These relationships of the Nar Valley Beds are such that, the palynological evidence being insufficiently precise, an Ipswichian date satisfies the facts as perceived. If the Nar Valley Clay is estuarine, then a sea-level up to at least 18 m O.D. within the Ipswichian must be postulated. That such a level does not accord with the presumed Ipswichian maximum suggests that, far from disregarding the Nar Valley evidence, a re-examination of the site is overdue.

Ipswichian sediments survive also in the Humberhead area (fig. 4.1), and contain micro-fossils suggestive of brackish water influence. Gaunt *et al.* (1974) have reviewed the significance of these and other interglacial deposits in relation to Ipswichian sea-levels. The temperate Austerfield, Westfield Farm and Langham deposits preclude a maximum mean sea-level higher than about one metre above present, which confirms the conclusions derived from the Sewerby cliff evidence. The sea-level problem will not be fully resolved, however, until the Kirmingham and Nar Valley sites have been carefully re-examined, and it seems wise here merely to conclude that sea-level was higher than present for much of the Late-Temperate stage of the interglacial, when the sea spread widely over the Fen basin and penetrated the Humberhead area at least at high tide.

During the fully vegetated temperate period, a complete vegetation cover and a predominance of chemical weathering would control runoff and sediment size in such a way that geomorphic activity by streams would be restricted mainly to flood-plain development with relatively fine-grained sediments. Most of the Ipswichian deposits rest on bedrock or pre-Wolstonian drift, and Gaunt *et al.* (1974) demonstrate incision to at least 13 m below O.D. prior to deposition of the Langham Ipswichian beds. Thus it would appear that the most opportune period for fluvial erosion was that immediately following Wolstonian deglaciation and preceding the establishment of a closed vegetation cover, namely the Late-glacial zone of the Wolstonian when glacifluvial and periglacial conditions may have prevailed, and part of the Pre-temperate zone of the Ipswichian. During the last, Post-temperate zone of the interglacial, as sea-level began to fall, fluvial erosion may well have increased again to reach optimum intensity in the Early Devensian.

Physiographically therefore it is realistic to regard the interglacial as longer than that period identified solely by temperate biostratigraphic zones, and as comprising an earlier phase of dominant fluvial erosion when valley gradients were steepest, vegetation cover was not closed, and sea-level was well below optimum, a middle period when deposition prevailed in the lower and possibly middle reaches of rivers in response to high sea-level, more regular river regimes and closed vegetation cover, and a later period when falling sea-level and increased climatic severity brought about renewed incision.

Some measure of the magnitude of valley incision for this longer period can be obtained for various parts of eastern England.

In the Lincolnshire Wolds, presumed Early Devensian tills lie in valleys up to 30 m below the base of Wolstonian tills. In Kesteven where valleys have been excavated by streams initiated on Wolstonian till which concealed former valley alignments, incision by most streams has been regionally of the order of 10-20 m. In parts of the lower Welland, Nene and Ouse valleys, acknowledging the variable height of the till base, up to 30 m of postglacial deepening has occurred. Though Sparks and West (1965) consider that relatively little post-Wolstonian erosion occurred in the Cambridge district, in west Norfolk the lower Wissey valley was deepened by some 15 m before deposition of Ipswichian sediments, and in northwest Norfolk (fig. 3.2) general dissection of Wolstonian tills prior to the emplacement of Devensian drifts was of the order of 20-30 m (Straw 1960). If the Lower Sand and Gravel (Gaunt *et al.* 1971) of the Aire valley is Early Devensian, an incision of over 30 m below nearby patches of Older Drift is indicated. The gravel deposits, probably glacifluvial, capping the isolated Triassic hills of Brayton Barff, Hambleton Hough and Holme-on-Spalding Moor, were considered by Bisat (in Melmore 1934) to have been laid down when the floor of the Vale of York was at least 30 m higher than at present.

This brief survey satisfies no purpose other than to demonstrate regional valley deepening of the order of 10-30 m as a measure of interglacial dissection. Denudation of pre-Ipswichian tills has been greatest in the northern hill areas and least in parts of East Anglia, and it may also be observed that these drifts are more widespread and better preserved in south-central Lincolnshire and the Fen basin than in the Humberhead, lower Trent and Vale of Belvoir lowlands.

The former presence of till over permeable rocks of the hill areas provides an additional mechanism for the formation of dry valleys. Most of these tills are rich in fine-grained materials and on withdrawal of Wolstonian ice would have constituted at first a relatively complete and impermeable cover to the Cretaceous, Jurassic, Triassic and Permian sandstones and limestones, encouraging surface runoff rather than infiltration. Eventually valleys developed on the till cover would be superimposed onto permeable bedrock, which process led to their eventual desiccation. Surface flow would continue in the upper parts of such valleys still developed in till, perhaps supporting streams in the lower reaches across the newly exposed permeable rock, but once the upper catchment areas were reduced below a critical size the streams would not be maintained and the systems would then become dry. Dry valleys formed in this way may well have been modified later by spring sapping or by periglacial niveo-fluvial processes.

The Ipswichian Interglacial, was, therefore, not an insignificant period in the physiographic evolution of eastern England. Further study should elucidate more details of fluvial and marine activity, but it was not a time when major changes of landform and drainage pattern were effected. In this respect it resembles the Devensian, and it is to earlier events still that the larger features of the landscape owe their origin.

5 The Wolstonian glaciation

In eastern England deposits of this glaciation survive mostly south of the Humber. Northward, only the Basement Till of Holderness, and vestigial patches of drift on the Yorkshire Wolds, in the northern and western Humberhead areas, and on the Magnesian Limestone cuesta are left as testimony of this event.

Westward of the Trent, small areas of gravel or till persist at Blidworth, southeast of Ollerton, at Whitwell and at Anston, where they spread onto Coal Measure rocks. Thin gravelly drift is present around the lower Idle valley at 30-65 m O.D., and outwash gravels survive on the narrow west-east Barrow Hills west of Gainsborough. In every instance, the drifts lie at the surface, occupy interfluve positions and are clearly remnants of formerly more widespread deposits. Of those areas where the Wolstonian drifts are more extensive it is most appropriate to discuss first that north and northwest of the Fen basin.

South Humberside and Lincolnshire were overwhelmed by ice which laid down lodgement tills distinctive in colour, composition and area of occurrence. The ice-sheet was composite, for while the Wolds were crossed by south-flowing North Sea ice, the Jurassic outcrops were traversed by 'inland' ice flowing southsoutheast from the Humberhead area (fig. 5.1). The junction between the two ice masses was largely along the Cretaceous scarp to Hainton, and then south to the Fen margin a few kilometres to the west of the Bain valley (Straw 1957). The bulk of the western drift has a fairly local provenance, and distinctive tills lie in contiguous parallel belts from west to east (Harmer 1928), proving the ice to have moved parallel to the strike of the Jurassic strata. East of Gainsborough the till incorporates much Keuper Marl, whereas south of the Ancaster gap

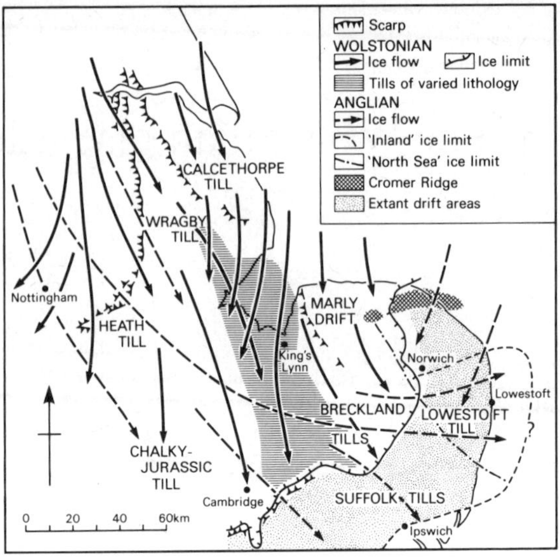

Fig. 5.1 Anglian and Wolstonian ice flows and limits over eastern England south of the Humber estuary.

it consists mainly of Lower Lias clays and limestones. East of Lincoln and Sleaford it contains more Middle Lias and Oolite material, and in the central Lincolnshire vale from Brigg to the Fens the blue-grey Wragby Till results from the reworking of Oxford, Corallian and Kimmeridge Clays (Straw 1969). Ipswichian sediments rest directly on this till beneath Devensian gravels at Tattershall (Girling 1974).

The characteristic deposit of the Wolds and the Bain valley area is a creamy-white to light-brown highly calcareous till (Calcethorpe Till) which still covers large interfluve areas in the central and southwest Wolds. On the west side of the Bain valley, it overlaps a grey and brown till dominated by Lower Cretaceous materials (Belmont Till) which outcrops narrowly along the valley-side for a few kilometres south of Biscathorpe (Straw 1966). In the eastern Wolds at Welton-le-Wold it is closely associated with a brownish-grey till (Welton Till) which has affinities with the Basement Till of Holderness (Catt and Penny 1966) and overlies flint valley-gravels containing a meagre probably Hoxnian fauna (Alabaster and Straw 1976).

It is important to stress this spatial distribution of the Wolstonian tills in Lincolnshire because it could not have been produced by ice which flowed orthogonally to bedrock outcrops. The inferred flow direction can be extrapolated reliably into areas further south and southeast, and as a consequence certain geomorphological features can be ascribed more positively to this glaciation (fig. 5.2). The disposition of the drifts in the middle Bain area (Straw 1966) suggests that the North Sea ice pushed over the southern half of the Wolds and probably through the Wash gap rather later than the occupation of the central vale by the inland ice (fig. 5.1).

Figure 4.2 shows the form of the subtill surface in central Lincolnshire. The cuesta and vale landscape of eastern England existed prior to glaciation, but substantial modifications to parts of it were effected during both the passage of ice and subsequent deglaciation (fig. 5.2). Over the limestone hills, ice-sheet erosion was of variable intensity. Erosion is indicated by the absence of

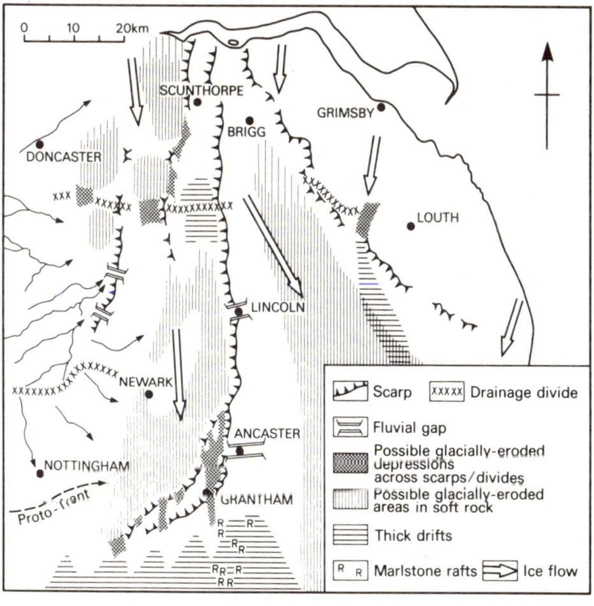

Fig. 5.2 Erosion and deposition by Wolstonian ice in the central part of eastern England.

subaerially weathered materials and the rarity of earlier drifts beneath the tills, by unweathered Jurassic limestone and sandstone, chalk and flint clasts in the various tills, and probably by the subdued form of Lincoln Edge, allowing for the northward attenuation of the Middle Jurassic rocks, and by the lack of preglacial planation surface remnants south to the Ancaster gap. Erosion of Upper Chalk, which is restricted to the northeast Wolds, is confirmed specifically by the distinctive tabular grey flints lodged in tills as far south as west Norfolk. Lower Chalk and Lower Cretaceous rocks were heavily eroded in the upper Bain area (Straw 1966). The local Belmont Till consists mostly of Lower Cretaceous materials, and whereas in pre-Wolstonian times a valley drained eastward from the Bain head area (Straw 1961a, 1976), the Bain now drains south, the former divide with a large obsequent valley having been grossly modified by the passage of ice (fig. 9.1).

Localized glacial erosion also occurred in southwest Lincolnshire. Harrod (1972) and Wyatt (1971) have located over twenty enormous erratic masses of 'rafts' of Middle Lias Marlstone up to 200 m across and several metres thick within the Wolstonian till (fig. 5.2). Such rafts must have come from the anterior scarp developed around Grantham. Wolstonian ice crossing the Vale of Belvoir would have overridden this scarp, quarrying and carrying away southward masses of Marlstone most probably from the sites of several anomalous gaps (Straw 1969). Similar circumstances account for the depth of the Witham valley not only where it is developed in Upper Lias clay north of Grantham, but also perhaps for a few kilometres south into the Lincolnshire Limestone to Great Ponton. Though the Witham valley here coincides with a preglacial depression (Wyatt 1971), modification by ice is indicated by a Lincolnshire Limestone raft (129 m long by 10 m thick) in till near Great Ponton (Morris 1853).

The valleys suffered greater general erosion (fig. 8.1), and up to 30 m of rock may have been removed from the Vale of Belvoir (Straw 1963a; Rice 1968), the lower Trent and Till basin northwest of Lincoln, and central Lincolnshire and the Fens (Straw 1958). Liassic material dominates the tills of southwest Lincolnshire, over 60 m thick in places, and must derive from the Vale of Belvoir. Similarly, the dark grey tills of central Lincolnshire, the Fen basin and Cambridgeshire were constituted from Upper Jurassic clays outcropping south of the Humber.

South of Scunthorpe, the form of the low-lying till-covered areas around Scotter and Blyton between the weak Lias scarp, the isolated eminence of Hardwick Hill in Scotter Common, and the small Rhaetic-Lias cuesta at Gainsborough suggests moulding by Wolstonian ice (fig. 5.2). Glacial erosion may even account for the Trent valley at Gainsborough, where it is notably constricted between the Rhaetic slopes and the eastern end of the Gringley Hills. The Trent did not follow this course in pre-Wolstonian times (Straw 1963a) and glacial breaching of the ridge is a feasible way by which the post-Wolstonian Trent gained eventual access to the Humber basin. At Balby near Doncaster, till survives in glacially-eroded depressions in Bunter Sandstone as low as 24 m below sea-level.

The survival of preglacial planation surface remnants in parts of the Lincolnshire Wolds and Kesteven demonstrates that ice erosion here was not excessive. The subdrift surface in Kesteven reveals that the tills infill and thereby conceal numerous east-draining pre-Wolstonian valleys as determined by many workers (Kellaway and Taylor 1952; Rice 1965; Wyatt 1971; Harrod 1972). Physiographic interest lies in the contrasting alignments of the present valleys with the subdrift systems (Wyatt 1971), in the spatial pattern of the surviving tills, up to 75 m in thickness, which conforms to a large extent with that of the buried valleys, and the subdued morphological expression of the till surface. The largest buried valley trends from the vicinity of Melton Mowbray through Castle Bytham to Thurlby, south of Bourne.

In the Castle Bytham area till overlies up to 18 m of sands and gravels which include Triassic as well as Jurassic material. If these sands and gravels are early Wolstonian, and the 1:1000 gradient of the bedrock floor of the valley represents the Hoxnian profile, then the 'hang' of 30 m of this floor at the Fen margin is a further measure of the bedrock erosion achieved by Wolstonian ice in the western Fen basin.

Wolstonian ice seems to have become stagnant over large areas of country south of the Humber during deglacierization, whence general thinning of the ice sheet led inevitably to the exposure of the hill areas before the vales. Progressive uncovering of the Kesteven plateau by Fen ice is recorded by the north-south oriented upper Witham, Glen and Eden valleys which denote a group of parallel ice-marginal channels now much deepened and modified by post-Wolstonian incision (fig. 9.1). On the eastern side of the Fen ice, the Bain valley and its lower left-hand tributaries record similar stages of shrinkage (Straw 1966). A residual mass in the Vale of Belvoir and west of Lincoln best accounts for the course of the renascent Trent across Trias rocks from near Long Eaton to Newark (Straw 1963a), while on its southeast side the river Witham utilizing the glacially deepened depression north and south of Grantham behind the Marlstone scarp, took up a course northward and then eastward through the Ancaster gap. Northward, integration of the Idle system west of the ice mass explains plausibly the abandonment of the Kirton and Eaton gaps across the Keuper Marl cuesta and on further contraction of the ice the Trent appears to have found its way north to Gainsborough, thereby extending its anomalous course across the dipslope of the Keuper Marl, and thence through the glacially-breached divide.

Within the Vale of Belvoir, the small river Devon, its tributaries and the Witham cross the various weak Liassic limestone ridges in a discordant manner that cannot be explained by conventional processes of spring-sapping and river capture. A mechanism involving superimposition of englacial and subglacial streams beneath a decaying ice mass satisfies not only these circumstances, but also those pertinent to the groups of streams in the central Lincolnshire vale which either flow normal to the gradient of the till surface, or have been let down across low ridges of Kellaways Rock and Cornbrash.

North of the Humber the effects of glaciation could have been of a similar order, but whether the Vale of York and the Humberhead area were eroded like the Vale of Belvoir and the northern Fen basin, and whether the river patterns of the eastern Pennines and the eastern uplands were modified during deglaciation will probably never be known. Too often, however, the glacial effects here are interpreted only in Devensian terms, and a longer glacial perspective is surely necessary.

West of the Fens the Kesteven tills extend south as a dissected sheet across the valleys of the Welland, Nene and Ouse systems. Recent work demonstrates the bulk of these drifts to comprise the Chalky boulder-clays, probably of Wolstonian date (Taylor 1963; Horton 1970).

Chalk-bearing tills also survive over the western plateau of Cambridgeshire at 67-82 m O.D. (Sparks 1957). As in Kesteven they present a monotonous surface but conceal a more complex preglacial relief, and overlie Upper Jurassic and Lower Cretaceous rocks. Over 60 m thick in places the till is generally bluish-grey in colour, often dark with fragments of chalk, flint and Jurassic rocks, Bunter pebbles and a few far-travelled erratics (Edmonds and Dinham 1965). A large erratic mass of Kimmeridge and Amptill Clay southeast of Biggleswade can only have been transported from the north or northnortheast. Norris (1962) regards the bulk of the till as of Gipping (Wolstonian) age, though Sparks and West (1965) describe the tills of the western plateau as of 'Lowestoft' type on the bases of composition and preferred stone orientation, and imply a pre-Wolstonian age. By contrast, although the Cam valley is deficient of till, Wolstonian ice moving south of the Fen basin must have entered it but probably the ice had little effect for, as argued below, it may not have surmounted the Suffolk escarpment.

In the Brecklands tills are discontinuous beneath the surface sands. Around Mildenhall the till is greyish-brown and contains both Jurassic and Cretaceous material. In other areas it is lighter in colour with sandy rather than clayey matrices, derived in part from the Sandringham Sands of west Norfolk (Perrin 1961). Holmes (1971) argues for a Saalian (Wolstonian) age and erratic suites suggest general movement of ice from the north and northnorthwest.

In west Norfolk two lithologically distinct tills are present. East and southeast of King's Lynn, a bluish-grey till incorporating Upper Jurassic and perhaps some Lower Cretaceous clays occurs sporadically on low hills and spurs of Lower Cretaceous rocks. In the Bawsey quarries, on top of Sandringham Sands, a highly chalky till is contorted and sheared with masses of dark grey till. Evans (1975) claims that the chalky till generally overlies the darker, but ascribes both to the same glaciation. Stevens (1959) recorded a blue-grey till beneath the Nar Valley Beds near East Winch, and a white sandy clay with chalk and flint about one kilometre to the northeast, and also referred them to the same glaciation. East of the insignificant Upper Cretaceous scarp, much of the chalk area of Norfolk is overspread by a partly dissected thin sheet of chalky till (Marly Drift) incorporating wave-rounded cobbles of flint, and tabular and black flints from the Upper Chalk which prove the parent ice to have come from the north. Lithologically and geomorphologically it is very similar to the Calcethorpe Till of the Lincolnshire Wolds (Straw 1967).

These brief descriptions of the tills surviving in areas adjacent to the Fen basin are intended to show that they could all have been emplaced by Wolstonian ice advancing from the north and northnorthwest (fig. 5.1). The carry of distinctive Lincolnshire rocks like the Elsham and Spilsby Sandstones and the Red Chalk into the Cambridge and Breckland areas is highly significant in this connection (Worssam and Taylor 1969). It is important now to recall the situation in Lincolnshire where North Sea ice was shown to have abutted inland ice, creating a zone west of the Bain valley where the very chalky Calcethorpe Till was deposited against and a little over the Belmont and Wragby Tills. The darker 'Jurassic' tills beneath and west of the Fens correlate directly with the Wragby, Heath and Kesteven Tills of Lincolnshire, whilst the chalky and sandy drifts of central and west Norfolk were surely emplaced by the North Sea ice sheet responsible for the Calcethorpe Till. It seems inevitable that between these till provinces as shown in fig. 5.1 the drifts in the King's Lynn-Ely-Cambridge-west Breckland area should exhibit mixed lithologies, because the North Sea ice which had crossed the Lincolnshire Wolds or had passed through the Wash gap had by then travelled at least 70 km over Upper Jurassic and Lower Cretaceous rocks.

The Suffolk tills also comprise lithological types differentiated by Baden-Powell (1948) into a darker (Lowestoft) type incorporating Jurassic clay and a paler (Gipping) one with a sandy chalky matrix. The former survives predominantly in northeast Suffolk around Lowestoft and most workers acknowledge an Anglian date. In addition to the Gipping valley tills Baden-Powell considered those of northwest Suffolk and west Norfolk to be of Gipping type. West and Donner (1956) supported the view that the ice responsible for the Gipping tills had indeed passed south over the higher ground of west Suffolk, but most later reconstructions ignore Baden-Powell's view that an eastern limit to this ice occurred in central East Anglia (Mitchell 1960; West 1968; Embleton and King 1968; Sparks and West 1972).

There is no valid reason for claiming that the ice which emplaced the dark tills in west Norfolk and Cambridgeshire was that which moved eastward from the Fens into east Suffolk to deposit the 'Jurassic' till around Lowestoft. To do so would perpetuate one of the basic fallacies that has become established in East Anglian stratigraphy since Baden-Powell's paper in 1948; specifically the view that *all* grey 'Jurassic' type tills in the Fen region and East Anglia were emplaced during the same glaciation. This assumption has led to correlation of the local Bawsey Till in west Norfolk (Baden-Powell 1948; Evans 1975), the Nar Valley Till (Stevens 1959) and tills of the western plateau of Cambridgeshire (West and Donner 1956) with the true Lowestoft Till of east Suffolk. Constrained by it, Baden-Powell was forced to dismiss the chalky drifts of central and northwest Norfolk as merely a different facies of the westerly-derived Lowestoft Till, thereby disregarding their contained evidence for southerly movement of the ice which so impressed him with respect to the Gipping-type tills elsewhere. The west Norfolk and Breckland tills are local variants produced by Wolstonian ice as it streamed south over the zone where Kimmeridge Clay, Lower Cretaceous and chalk outcrops are closely disposed. West and Donner's (1956) conclusions regarding regional ice-movement, based solely on till fabric analysis, bear re-examination, for preferred orientations reflect the influence of factors other than just direction of ice flow. A further simple but significant point is that if the chalk-bearing Jurassic-type tills of west Norfolk, Cambridgeshire and Huntingdonshire were emplaced by ice moving from the northwest, it is difficult to explain where the chalk came from.

The eastern limit of Wolstonian ice in Norfolk (fig. 5.1) can be traced west and northwest of Norwich where the Marly Drift terminates within a mass of outwash materials, designated the Telegraph Hill-Aylsham Gravels (Straw 1973). Further north Wolstonian ice overrode the Cromer Ridge westward of Edgefield and was perhaps responsible for the broad gap at Briston, but did not surmount the main part of the Ridge even though this barely exceeds 92 m O.D., and its limit therefore runs northeast to cross the coast near Weybourne. Eastward of Weybourne the Marly Drift gradually gives way to older glacial deposits over a terminal zone where the Marly Drift ice was pressing against the rising slopes of the Cromer Ridge, where some contortion with the older drifts probably took place, and where excessive valley development, slope recession and marine erosion have since occurred (fig. 7.1).

South of Norwich, the limit of the Wolstonian ice is less easy to detect, for chalky boulder-clays are more widespread. Cox and Nickless (1972) carry the ice-margin northeastward to the coast near Winterton, but it can be argued that the terminal complex of Wolstonian drifts recognized northwest of Norwich extends southward (Straw 1974). The generally lighter-coloured grey and brownish tills of central Norfolk which retain characteristics of the Marly Drift, give way east of the river Tas to darker Jurassic-type tills which then generally underlie the surface eastward to the coast, to be exposed in cliff sections at Scratby and Corton. The river Tas may therefore be successor to marginal or submarginal meltwater streams which at the maximum extent of the Wolstonian ice flowed northward to join those descending the Yare and Wensum valleys (fig. 9.1). An efficient single system of drainage along this section of the ice-front would explain the absence of a tract of outwash gravel such as occurs northnorthwest of Norwich (Straw 1973). North and northeast of Diss small, poorly-integrated streams occupy interconnecting shallow depressions on the till surface, features which may have arisen during the decay of a stagnant marginal zone of ice. A larger stream drains east from this area through Starston to join the Waveney at Wortwell, following a line of meltwater drainage which contributed materials to the late Wolstonian outwash sediments now constituting the main Homersfield terrace (Sparks and West 1968) of the Waveney. The Wolstonian limit may be indicated by the westward termination of this terrace near Diss. Six kilometres eastsoutheast of Diss, south of the Waveney, is the important interglacial site of Hoxne. Recent excavations have shown that the interglacial sediments are overlain by fluvial sands and gravels and a strongly cryoturbated 'head', and were not overridden by Wolstonian ice (Gladfelter 1975). The fluvial deposits may well be Wolstonian outwash materials carried northeast to the Waveney along the line of the river Dove before deeper incision of the Dove and Waveney valleys occurred. The river Dove drains an extensive area southsouthwest of Eye, and has a relationship to the Waveney similar to that of the Tas to the Yare.

The absence of a terminal moraine feature south of Norwich requires explanation, and the most plausible suggestion is that the ice-sheet had gradually lost momentum across central Norfolk until at the limit identified above, it had virtually ceased flowing, and the marginal zone quickly stagnated. Decay of the dead ice produced meltwater systems beneath and eastward of this zone which initiated the present drainage pattern, including even the upper Waveney and the upper Yare (fig. 9.1).

Recent studies of the Suffolk drifts (Bristow and Cox 1973) confirm only one suite of glacial drifts, even though more than one till may occur locally, as for example in the Gipping valley. The age of this till complex is not agreed, however, for Bristow and Cox prefer a Wolstonian but pre-Hoxnian age, thereby questioning the status of the Hoxne interglacial site, whereas others (Bristow and Cox

1973, discussion; Turner 1970) offer a more orthodox argument favouring an Anglian age and deny the occurrence in Suffolk of Wolstonian ice. The behaviour of the Wolstonian ice at and after the time of its postulated maximum coverage of Norfolk carried however two implications regarding its possible extent in Suffolk. First, if the ice movement over south Norfolk was very weak, then it is difficult to envisage it surmounting the chalk scarp of northwest Suffolk which reaches about 120m O.D. Second, the curve of the postulated ice margin from Norwich to Diss extrapolates toward Elmswell and Bury St Edmunds, south of which the ground rises generally to the High Suffolk plateau around 90m O.D.

The dissected gravel spreads about the several headstreams of the river Gipping between Bury and Wetherden assume significance in this connection, for the Gipping valley is the only Suffolk valley, south of the Waveney, which could have carried Wolstonian meltwaters directly away from the ice front.

Limited penetration by Wolstonian ice into Suffolk explains in part the contrasts in landform and glacial stratigraphy that exist between the High Suffolk plateau and the Brecklands. The latter are everywhere less than 60m O.D. and along their southern margin between Thetford and Mildenhall, isolated deposits of lacustrine sediments survive both within and above glacial tills, and sites near Elvedon (Patterson and Fagg 1940) and Mildenhall (Holmes 1971) have yielded large numbers of Acheulean artefacts. The characteristics of the drifts and the formation on deglaciation of the irregular surface morphology indicated by the lacustrine hollows and gravel mounds can be explained satisfactorily in terms of an oscillating margin of a thin ice-sheet that eventually stagnated and decayed over at least parts of the southern Breckland. Holmes (1971) ascribes the drifts near Mildenhall to at least two ice advances within the Saalian (Wolstonian) Glaciation and points out that the glacial drifts within the present valleys of the Lark, Kennet, Snail and Cam are overlain only by fluvial terrace gravels and are affected only by Devensian periglacial structures. A Wolstonian age is therefore most likely (table 6.1) and it seems reasonable to extend the southern limit of the Wolstonian ice along the rising chalk slopes westward from Bury St Edmunds to Newmarket, and thence southwest along the scarp toward the Gog Magog Hills and the Hitchen gap (fig. 5.1).

South of the latitude of the Wash, the subdrift surface is one of considerable relief and composite form. Most of this surface is probably little modified, but localized subglacial erosion did take place because elongate and linear depressions of an overdeepened character have been described within the Ouse, Nene and Welland basins (Horton 1970; Wyatt 1971) and from central and west Norfolk and the Breckland (Woodland 1970). The composition of the Wolstonian tills and absence of palaeosols and weathered bedrock beneath them point to general if slight erosion of both Cretaceous and Jurassic outcrops as in Lincolnshire.

Physiographically, the main consequence of Wolstonian glaciation in Norfolk, Cambridgeshire and Huntingdonshire was the deposition of till thick enough in many places to obliterate pre-existing valley systems, but neither glacial erosion nor deposition was of a magnitude sufficient to modify grossly the major features related to the solid geology. However, the physical consequences of Wolstonian deglaciation in relation to drainage systems are more obvious, for all the major drainage elements were established or re-established as the ice disappeared, and have not been disrupted since by Devensian ice (fig. 9.1).

Woodland (1970) showed that parts of the present-day valleys accord closely in their alignment with subglacial depressions. This is partly true of the river Wensum which is successor to subaerial meltwater streams which coursed generally eastward across the newly exposed surface of the Marly Drift (Straw 1973). These streams deposited several convergent gravel trains (Wensum Gravels) which persist as dissected high-level terraces bordering the Wensum and some of its tributaries. In northwest Norfolk these gravels pass headward into outwash plains around Bircham, broadly contemporary with those around Holt and Kelling in north Norfolk. The Yare, Tas and Upper Waveney also exhibit some relations to buried channels, but excavation of the present valleys was begun by subaerial meltwater streams flowing from the westerly-receding ice margin across the progressively-exposed drift surface.

Contrary to Woodland's (1970) view, the Wolstonian ice fronts in Norfolk and the Breckland withdrew toward the North Sea and the Fens and at a late stage streams were re-established draining toward rather than away from them. The head of the Nar valley is of particular interest here (Straw 1973). Above West Lexham the stream, flowing west, is incised below and post-dates east-grading Wolstonian outwash gravels (Wensum Gravels). A low terrace borders the Nar, and in the Narborough area lies at heights comparable to the interglacial Nar Valley Beds, placing in question again therefore the postulated Hoxnian age of these beds (Stevens 1959).

The Breckland streams also drain to the Fens, and are bordered by spreads of fluvial sands and gravels that constitute up to four distinct terraces (Worssam and Taylor 1969). Ipswichian interglacial sediments occur in the Wissey valley (Sparks and West 1970), and as the terraces are clearly post-Wolstonian the present pattern of streams is surely no older than Late Wolstonian. Except for the Lark, the valleys are sinuous in plan, almost casual in their arrangement, and in places are connected by shallow depressions as, for example, those between the Wissey and the Thet systems east from the Ickburgh and Little Cressingham. Such poorly organized drainage is consistent, accepting some influence by the form of the subdrift surface, with the gradual development of the stream pattern within, beneath and around decaying dead ice, a deglacial condition already inferred from the stratigraphic and morphologic features of the southwest Breckland for the Wolstonian ice.

Southsoutheast of Cambridge, possible Wolstonian effects on drainage relate to the upper Cam near Saffron Walden. Within a shallow scarp-crest col, the upper Cam drains north against the chalk dip from near Quendon in Essex, coincident with a drift-filled 'tunnel-valley', to reach the Lower Chalk plain through an escarpment re-entrant near Chesterford. However, its tributary pattern suggests former southerly drainage to the river Stort, and south of Quendon, broad spreads of gravel around Ugley and Stansted Mountfitchet descend into the Stort valley. The 'tunnel-valley' and the sandur-like gravels around Ugley suggest southward, not northward (Woodland 1970) flow of subglacial waters beneath Wolstonian ice, and though it has been argued that this ice-sheet failed generally to override the chalk scarp, a marginal lobe could have pushed into the existing col as far as Quendon, eventually to discharge meltwaters down the Stort valley.

6 The Hoxnian interglacial

Because this interglacial is more distant in time from the present and fewer sediments and morphological features have survived, reconstruction of environmental conditions and physical events is more difficult. Biological evidence from the type locality at Hoxne in Suffolk (West 1956), and from Marks Tey (Turner 1970) and Clacton (Pike and Godwin 1953) in Essex is sufficiently strong however both to differentiate this interglacial from that of the Ipswichian and to effect correlation with the continental Holstein interglacial. Turner (1970) suggests that the interglacial *sensu stricto* lasted approximately 30000 years. The available evidence points to an amelioration of climate to a fully temperate condition with re-establishment of a rich flora and fauna through major stages comparable with those of the Ipswichian interglacial.

Bristow and Cox (1973) have argued conversely that the Hoxnian and Ipswichian interglacials are separated only by a nonglacial cold phase, and that both post-date the Wolstonian Glaciation. These views stem from their belief that the tills of East Anglia belong to one glaciation, allied to the incontrovertible fact that the Hoxne interglacial deposits rest on an undoubted chalky boulder-clay. A case has already been presented for Wolstonian glaciation of part of East Anglia and as the biological evidence for two discrete interglacials appears overwhelming, the more orthodox interpretation is preferred here.

Intensified valley erosion toward the end of the interglacial is confirmed by evidence from Lincolnshire. Frequent occurrences of solifluction and scree gravels are known in various buried valleys beneath chalky boulder-clay (Wyatt, 1971; Wyatt *et al.* 1971), but the Melton Mowbray-Thurlby channel containing up to 18 m of fluvially-transported sands and gravels provides the major instance. At Welton-le-Wold in the eastern Lincolnshire Wolds, some 13 m of Wolstonian tills (Welton and Calcethorpe Tills) overlie at least 8 m of flinty valley-gravels and sands (Welton Gravels). The gravels accumulated under periglacial conditions, enclose frequent lenses and seams of windblown silt, and have yielded a sparse intrusive mammalian fauna of Hoxnian character and a few Acheulean handaxes (Alabaster and Straw 1976). The 'valley-gravels' include Lower Cretaceous materials and were deposited along the lower section of a major Wold valley which, during Hoxnian times, drained eastward from the western Wolds. The head of this valley was grossly modified by Wolstonian ice and now lies within the upper Bain catchment.

Most of the reported Hoxnian interglacial sediments (fig. 4.1) were deposited in freshwater conditions, but traces of marine influence at Clacton (Pike and Godwin 1953) suggest a sea-level up to 9 m O.D. in the Late-temperate (Ho III) zone.

The Nar valley and Kirmington deposits contain estuarine deposits indicative of high-tide levels around 20-23 m O.D., but the true age of the sediments at both localities is still in doubt. On the north side of Flamborough Head the Upper Speeton Shell Bed, consisting of some 3 m of silts and sands with estuarine molluscs, is probably Hoxnian and reaches 32 m O.D. (Penny 1974). West (1972) deduced a sea-level of at least 23 m O.D. for the latter half of the interglacial, but the location of a Hoxnian shoreline at this height in eastern England is difficult to determine. Hoxnian sediments have not been discovered to date west of the Yorkshire and Lincolnshire chalk, probably a consequence of destructive Wolstonian and Devensian glacial and periglacial activity. But if, as implied in chapter 4, the clay vale floors in these areas were not much lower than about 30 m O.D. before Wolstonian glaciation, then the Hoxnian sea may never have been able to flood areas west of the chalk, and its penetration of the Fens may have been restricted to the vicinity of a narrower Wash gap. If these circumstances did prevail then the Hoxnian shoreline probably extended less far inland of the modern coast than did the Ipswichian. In Holderness the marine plane beneath the Basement Till affirms transgression during the interglacial, and would have been backed by cliffs. The Ipswichian sea subsequently planed the Basement Till and displaced the chalk cliffs to their present position. Similar circumstances probably occurred south of the Humber except that the Hoxnian cliff might survive as a low bluff backing a fragment of platform at 28 m O.D. south of Barrow-on-Humber (Straw 1964).

In Suffolk, the Hoxnian sea reached neither Hoxne nor St Cross South Elmham (West 1961), and neither deposits nor planed unconformities have been identified in the cliff sections along the east coast.

It may be claimed therefore that the highest presumed Hoxnian shoreline, predominantly cliffed, lay seaward of the modern East Anglian coast, extended minimally into the eastern Fens, closely paralleled the buried cliff of East Lincolnshire and Holderness and, like the Ipswichian shoreline, lay close to the modern coast north of Flamborough Head.

The scale and pattern of dissection of eastern England during the Hoxnian Interglacial might be assessed broadly by examining the distribution of drifts and by restoring the Wolstonian modifications to the drainage systems. The Wolstonian drifts, like the Devensian, rest by and large on a bedrock surface, and there are two possible explanations why this is so. First, interglacial denudation may have been efficient enough to remove pre-existing tills before deposition of later ones. Second, the later ice-sheet may have incorporated and reworked most of the earlier drifts within its area of occurrence. The efficacy of interglacial denudation alone can be challenged directly, for if it was responsible for removing Anglian till from Lincolnshire and west Norfolk prior to Wolstonian glaciation, why was more not removed from Suffolk? Also, if Ipswichian denudation had cleared the Wolstonian drifts from the areas of Devensian glaciation, why do such drifts survive immediately south of the Devensian limit? Thus ice-sheet erosion would seem to be the more important regionally in destroying earlier drifts, and in this connection it is significant that both Devensian and Wolstonian tills rest generally on unweathered bedrock surfaces and that there is a dearth of older interglacial sediments within a particular glacial province (Wymer and Straw 1977). Extension of this argument that ice-sheets largely destroy earlier drifts, to Suffolk and east Norfolk where Anglian drifts and Hoxnian interglacial deposits survive, supports the view that these areas were not glaciated during the Wolstonian.

North of the Humber there is some indication that during the Hoxnian interglacial, the Pennine rivers entered a Trias vale between 25-40 m O.D. and reached the sea variously by way of the lower Tees, Vale of Pickering and the Humber gap (fig. 9.1). South of the

Humber, the Lincoln gap was occupied by a proto-Witham, whose headstreams on the evidence of the aligned rivers Maun, Meden, Poulter and Ryton, and of the Kirton and Eaton cols across the Keuper Marl at 53 m and 42 m O.D. respectively, were located on Triassic and Permian rocks near Mansfield.

Southward, the Trent flowed east from Long Eaton through the Ancaster gap and across the Upper Jurassic outcrops to the sea in the vicinity of the Wash (Straw 1963a). The highest point of the rock-floor of the Ancaster gap is 38 m O.D., and in another section of the Hoxnian Trent valley west of Cropwell Butler the ground lies at 45-53 m O.D. Thus the Trent probably crossed what is now the Vale of Belvoir at about 40-45 m O.D., with rather higher ground to the north on the divide with the Hoxnian Witham. Prominent benches in the Ancaster gap at 55-70 m O.D. were claimed by Harrod (1972) to mark the height of the Hoxnian Trent, but this altitude is difficult to square with the 30 m height of the nearby Thurlby buried channel at the Fen margin. This channel and the other buried valley systems of the Middle Jurassic cuesta west of the Fens are clearly manifestations of subaerial dissection during the Hoxnian interglacial.

Within those parts of Norfolk that were covered by Wolstonian ice Hoxnian drainage lines are difficult to determine but some of the tunnel-valleys identified by Woodland (1970) were probably excavated along pre-existing depressions, for example that beneath the Wensum below Worthing. Broader, shallower sub-drift depressions may indicate the main elements of a proto-Wensum system that flowed generally southeast from northwest Norfolk (Straw 1973). In east Norfolk the extant Bure-Ant system was not disrupted directly by Wolstonian ice. These rivers and their tributaries rise on the southerly slopes of the Cromer Ridge, likely successors to outwash streams from receding Anglian ice. Throughout the Hoxnian interglacial, therefore, they flowed southsoutheast eroding into outwash deposits, Norwich Brickearth and Crag, eventually to meet the early Wensum in the Reedham area southwest of Great Yarmouth.

Coastal recession may well have destroyed an eastern segment of the Bure-Ant system. There is every indication that the Cromer Ridge formerly extended well beyond the present coast, and other south-flowing streams should have existed (fig. 6.1). One such is indicated by the small stream that now descends sharply to the sea at Mundesley, though upstream it flows southeast in a fairly broad valley, floored by highly fossiliferous Mundesley River Bed deposits

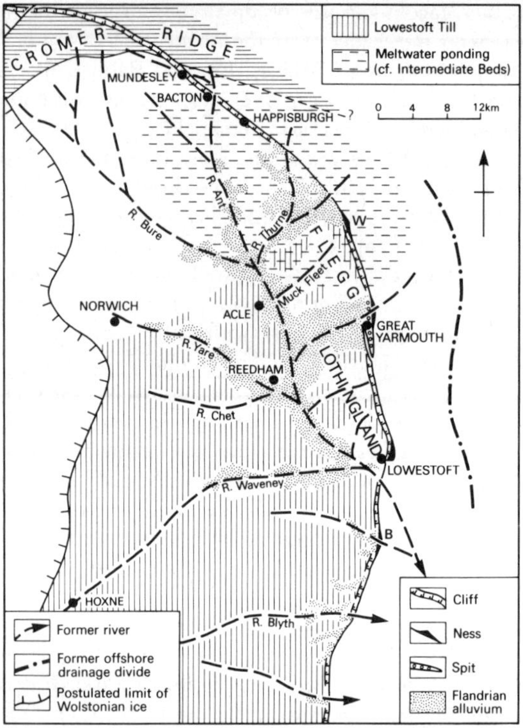

Fig. 6.1 Some physical features of eastern East Anglia, and a reconstruction of the Bure drainage system for the Hoxnian interglacial (W — Winterton Ness; B — Benacre Ness).

(Reid 1882). Four kilometres southeastward along the coast at Bacton, some 3-10 m of fluvial terrace deposits (Bacton Valley Beds) are at present exposed along the top of a 400-metre-long section of the low cliffs, inland and south of which a broad alluvial flat merges with that of the river Ant near East Ruston. Marine erosion has therefore dislocated the Mundesley-Bacton Valley, probably, accepting West's (1961) suggested date for the Mundesley River Bed, late in the Ipswichian Interglacial.

Southeast of the Ant, the river Thurne constitutes the remnant of an even larger left-bank tributary to the Bure. Though the latter might once have reached the sea by way of the Thurne marshes, it can be argued more easily that these marshes occupy a marine-enlarged basin that developed after coastal recession had consumed not only the headward areas of a longer Thurne valley but also most of its eastern flank (fig. 6.1).

West of Great Yarmouth, the rivers Bure, Yare and Waveney converge on the extensive marshes around Bredon Water, the Waveney avoiding a direct outlet to the sea through Oulton Broad at Lowestoft. If the Bure and Yare ever flowed southeast of Acle and Reedham respectively and conjoined as their general alignments indicate, then the trunk river should have crossed the modern coast over Oulton Broad. In the manner of the Thurne and the valleys northwest of Lowestoft another tributary may have joined this proto-Bure from the northeast through Yarmouth, draining an area between Flegg and Lothingland 'islands'. Marine erosion breaching the head of this valley in either the Hoxnian or the Ipswichian Interglacials provided a new route to the sea, since greatly deepened and enlarged. This tentative reconstruction demonstrates that Hoxnian river systems existed seaward of the modern coastline and that interpretations of the landforms, drainage systems and even the glacial stratigraphy of eastern East Anglia must acknowledge the great cumulative loss of land by marine erosion through the several interglacials.

The rivers which drain High Suffolk were resuscitated on Anglian glacial materials, but this is not to deny that their valley alignments were influenced strongly by subdrift depressions including tunnel-valleys and preglacial valleys (Boswell 1913; Woodland 1970). Boswell pointed out the general accordance of the valleys with the dip of the chalk and even suggested that certain inherited 'strike' sections were related to preglacial captures near an Eocene scarp. What is certain is that subaerial drainage was revived as Anglian ice withdrew, meltwaters commencing incision into the drifts to initiate a period of valley development that has proceeded not only through the Hoxnian interglacial but without further glacial interference to the present day.

Table 6.1 Correlation of some Middle and Upper Pleistocene deposits and events

Pleistocene stages		*Climatic regime*		*East Anglia*
DEVENSIAN	Late	RECURRENTLY PERIGLACIAL	GLACIAL	Breckland sands Docking Common Upper Gravel
	Middle			Docking Common palaeosol
	Early		GLACIAL	Docking Common Lower Gravel Wretton upper-most gravel and sand *Heacham advance* (Hunstanton Till)
				Wretton gravels (Wissey)
IPSWICHIAN		temperate		Morston 'raised beach' Interglacial deposits (Bobbitshole, Wortwell, Mundesley, Beetley, Swanton Morley, Wretton, ?Nar valley)
WOLSTONIAN		PERIGLACIAL	GLACIAL	Holt/Bircham outwash gravels Wensum Gravels (Wensum) Homersfield Terrace gravels (Waveney) Telegraph Hill/Aylsham Gravels *Norwich/Breckland advance* (Marly Drift, Breckland tills, west Cambridgeshire tills)
HOXNIAN		temperate		Interglacial deposits (Hoxne, St. Cross South Elmham, ? Nar valley) Proto-Bure drainage system
ANGLIAN		PERIGLACIAL		Briton's Lane/Roughton outwash gravels
			GLACIAL	*Cromer Ridge readvance* (Higher Cromer tills and gravels) Intermediate Beds *Lowestoft advance* (Lowestoft Till, Suffolk tills) Corton Beds *Norwich/Kessingland advance* (North Sea Drift, Norwich Brickearth, lower Cromer tills)

Lincolnshire and the Fens	*Yorkshire*
Cover sands Floodplain Terrace gravels – *Lake Humber II* – (Trent) *Hogsthorpe/Killingholme readvance* ––– (Upper Marsh Till, 'Marine Gravels') First Terrace gravels (Cam)	Cover Sands *Lake Pickering II* – Seamer gravels '25-foot Drift' Middle Silt (Aire) *Cayton/Speeton, York/Escrick readvances* (Drab/Purple Tills, Kelsey Hill/ Keyingham gravels)
Kirkby Terrace gravels (Bain) Tattershall fan gravels (Bain) Fen border gravels (Welland, Nene, Ouse, Wissey, Nar) Second Terrace gravels (Cam)	Middle Sand and Gravel (Aire) Lower Silt (Aire)
Beeston Terrace gravels – *Lake Humber I* – (lower Trent) –– *Lake Fenland* – Kirkby Moor Sands, Hemingby Terrace gravels (Bain) ––– *Stickney/Horkstow advance* ––––––– (Lower Marsh Till)	*Lake Pickering I* – Ayton/ Wykeham kame terrace Older Littoral Sand and Gravel Lower Sand and Gravel (Aire) *Melton, Wykeham, High Hutton, Stutton advance* Older River Sand and Gravel (Idle/Don)
Third Terrace gravels (Cam) March Gravels Fen Margin Gravel Interglacial deposits (Tattershall, ? Kirmington)	Older River Sand and Gravel (Idle/Don) Sewerby 'raised beach' Lower Speeton shell bed Interglacial deposits (Austerfield, Langham, Westfield)
Total glacierization (Welton Till, Calcethorpe Till, Biscathorpe Gravels, Wragby Till, Heath and Kesteven tills) Welton Gravels Bytham/Thurlby gravels	*Total glacierization* (Basement Till)
Interglacial deposit (? Kirmington) Brocklesby Park Surface	Interglacial deposit (Upper Speeton shell bed)

7 The Anglian glaciation

The deposits at Hoxne and St Cross South Elmham in Suffolk, and at Marks Tey in Essex rest unequivocally on chalky boulder-clay, proof that the most important and extensive glaciation of East Anglia, aptly called the Anglian, preceded the Hoxnian Interglacial. All glacial drifts outside the Wolstonian limit as defined above are here referred to this glaciation, and are specifically the North Sea Drift of east Norfolk, the true 'Lowestoft' Till of southeast Norfolk and east Suffolk, and the chalky boulder-clays of High Suffolk and Essex (fig. 5.1). The physiographic significance of the deposition of thick glacigenic materials over this vast area is great, for these drifts provided an early Hoxnian landscape very different in character from the preglacial one.

In Jurassic areas west of the Fens, fragmentary drifts of doubtful age, containing Bunter quartzite pebbles but no chalk and flint, survive in slight depressions beneath Wolstonian tills (West and Donner 1956; Posnansky 1960; Taylor 1963; Harrod 1972). Though southern Pennine ice probably crossed the Trent valley in the early Wolstonian, it could not have spread far over the Jurassic cuesta before the main incursion of northeastern ice (Posnansky 1960; Bridger 1976). However, Baden-Powell (1948) considered Pennine ice to be the upstream part of that which deposited his Lowestoft Till and there seems to be no reason why ice from the western Humberhead area should not have tracked from the Coal Measure and Permo-Triassic areas of east Derbyshire and north Nottinghamshire over the Lias and Oolite outcrops of southwest Lincolnshire. If the Hoxnian Trent crossed the Vale of Belvoir at about 40-50 m O.D. to occupy the Ancaster gap, then the higher benches within the gap at 55-70 m O.D. may represent the gap floor in pre-Anglian times when the Middle Jurassic scarp would already have been quite a prominent feature. It is likely therefore that the Anglian ice invested a landscape of considerable relief. A preglacial Cretaceous scarp was certainly in existence by the onset of Anglian glaciation (Straw 1970), but its alignment parallel to the direction of ice movement probably ensured little modification compared with that of the Jurassic cuesta north of the Ancaster gap, across which Anglian ice flowed strongly if obliquely. Further north, the North York Moors was undoubtedly an upland area by Anglian times, but whether it was then overwhelmed by ice is not known.

The general stratigraphy of the Anglian deposits (West 1968; table 6.1) is revealed at Corton, north of Lowestoft, in cliff sections where grey chalky boulder-clays (Lowestoft Till) overlie a series of sands and gravels (Corton Beds) and a brown sandy till (North Sea Drift). The latter rests unconformably on grey silts, representative of the pre-Anglian Cromer Forest Bed Series. The uppermost till, in places doubly layered with intervening outwash sediments (Banham 1971) is bluish-grey in colour and contains much Jurassic and some Triassic as well as Cretaceous material. It underlies the Hoxne interglacial deposits, and Harmer (1928) showed its distribution to extend approximately from the Gipping valley, eastward of the Breckland and the Tas valley, north toward the Yare. East of Norwich the till is rather lighter in colour, but in cliff exposures at Scratby, to the north of Great Yarmouth, it bears close lithological resemblance to the till at Corton. This distribution, precisely outside the Wolstonian limit postulated above, is a direct consequence of ice flow eastward from the southern Fen basin (fig. 5.1) which took advantage of the broad Lopham depression across the East Anglian chalk (Straw 1970).

The Corton Beds are water-lain sands, usually 10-15 m thickness, and contain a large marine molluscan fauna and occasional seams of clay, one such yielding a cold flora (West and Wilson 1968). Although previously mapped extensively in east Norfolk, their exact lateral correlatives are not in fact well known, especially along the coast (West and Banham 1968; Banham and Ranson 1970). They accumulated during an ice-free period, perhaps in an alternating freshwater and marine environment.

The brown sandy North Sea Drift contains some Scandinavian as well as north British erratics and has been correlated variously with the Norwich Brickearth and the Cromer Tills of east Norfolk. It represents an ice advance from the north to northeast Suffolk east of a line from Sheringham to Norwich and Kessingland (fig. 5.1). In the Cromer, Mundesley and Happisburgh areas three separate grey tills have been identified (Reid 1882; Banham 1968) with intervening sands and gravels, but southward and westward the Norwich Brickearth (browner, sandier and frequently decalcified) is less complex. Partly a shallow-water deposit, its almost horizontal disposition, disregarding later dissection, contributes much to the low plateau aspect of the ground northeast of Norwich.

The tripartite succession at Corton represents therefore three major events within the Anglian Glaciation of East Anglia. First was the ice advance from the north which emplaced the oldest tills in Britain over a variety of Middle Pleistocene estuarine, marine and fluviatile deposits, but which did not carry into central or west Norfolk nor over Suffolk. Second was the oscillatory withdrawal of this ice, an extension of flooding (fig. 6.1) associated with the Norwich Brickearth, and deposition of the sandy Corton Beds in the south and probably one or more of the outwash layers interleaved with the Cromer Tills in east Norfolk. Third was the flux of western ice over south Norfolk and northeast Suffolk, displacing the water and depositing thick impermeable tills over both outwash materials and Corton Beds which so raised the land surface that it has never subsequently been submerged.

This third event was not confined to the east central area of East Anglia, for the tills of High Suffolk were most likely emplaced at this time. Near the Gipping valley the grey Lowestoft Till passes into more chalky tills, which have however been ascribed a later date by some workers (Baden-Powell 1948). Spatially the Suffolk tills seem to range unsystematically from lighter-coloured chalkier tills to darker varieties containing more Jurassic debris. Most likely these are facies changes, the lithological variability arising from the fact that the ice first crossed the Jurassic outcrops of Cambridgeshire and Huntingdonshire before overriding the chalk, erosion of which and intermixture with the Jurassic material would not necessarily be uniform.

However, over the southeastern half of the Suffolk chalk the tills rest on outwash sands and gravels and early Tertiary rocks which now outcrop on the flanks of valleys that dissect the till. Clearly no incorporation of chalk into the till occurred in this zone, but over the higher areas of the chalk cuesta between Bury St Edmunds and

Saffron Walden, where the till rests directly on bedrock, erosion by the ice-sheet must have occurred. If the Wolstonian ice did fail to overrun the scarp, then displacement several kilometres down-dip from the base of the chalk of both the crestline and scarp (Embleton and King 1968) must be ascribed to Anglian ice erosion. This displacement stands in marked contrast to the Chilterns scarp southwest of Hitchin which was not crossed by Anglian ice. Subglacial scarp recession goes a long way toward explaining the extent of the chalk plain of the Cam valley, and casts serious doubts on Sparks' claim (1957) that preglacial erosion terraces survive along the scarp face.

The tills average 30-50 m in thickness over High Suffolk, disposed generally in one relatively simple layer over the interfluves, whereas several tills with pronounced structures frequently characterize the major valleys. The latter are at least in part coincident with 'tunnel-valleys' which terminating near to the southeastern limit of the till must be Anglian, not Wolstonian, subglacial features (Woodland 1970).

Inland from Mundesley and Cromer, the North Sea Drift thickens excessively to form the bulk of the Cromer Ridge, the dominant relief feature of north Norfolk. From the gently undulating crestline, which reaches about 92 m O.D. along some 14 km, the ground descends gradually south within the Bure-Ant basin along interfluves that carry outwash sands and gravels, for example around North Walsham. By contrast the northern flank of the Ridge is steeper and more irregular (fig. 7.1).

The coast sections between Cromer and Mundesley reveal the structures of the Ridge to be of two types (Banham 1975), primary structures developed during deposition of the tills, and secondary large-scale diapirs resulting from load compaction. The former incorporate large imbricated thrustbounded 'rafts' of chalk and Middle Pleistocene sediments that are to be seen in cliffs at Trimingham, Sidestrand and between East and West Runton. In spite of the magnitude of these contortions, the present morphology of the Ridge along and south of its crestline comprises gentle slopes which reflect its physiographic antiquity (Straw 1965). The more pronounced relief features of the northern flank of the Ridge are not, contrary to earlier opinion (Harmer 1928), a primary morphological expression of the underlying structures but as shown in fig. 7.1 are a product of later valley erosion. Controversy persists as to

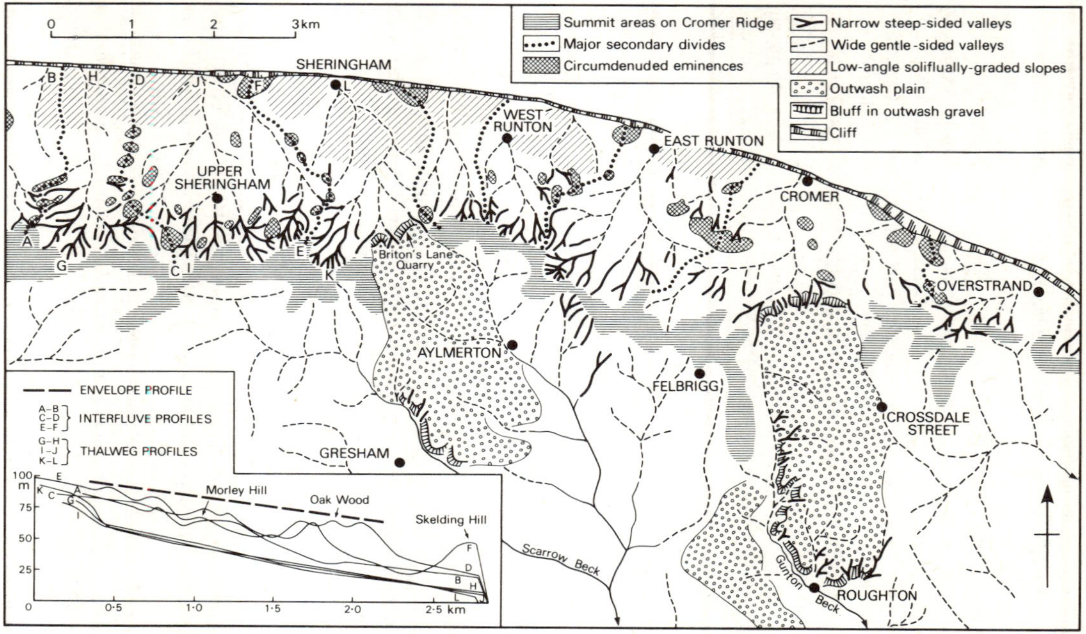

Fig. 7.1 Some physical features of the Cromer Ridge around Cromer and Sheringham.

whether this drift complex was formed beneath North Sea ice during its advance to Suffolk (Banham 1975) or whether it was built up as a 'push moraine' at a later stage when the North Sea ice either halted at or readvanced to this line across north Norfolk, penecontemporaneously with the sweep of western ice across Suffolk. Whatever way the Ridge was formed, it is certain that it formerly extended beyond the modern coast, and that ice lay along its northern face for some time during deglacierization. To speculate grossly, if the inland ice flowing toward the northeast beyond Scratby had approached the Ridge some 10-15 km east of Happisburgh (fig. 6.1), the resultant ponding of meltwaters over east Norfolk would provide the appropriate depositional environment for the Intermediate Beds, and the Mundesley and Gimingham Sands (West and Banham 1968; Banham and Ranson 1970).

Final withdrawal of both the northern and western ice permitted the establishment of the systems of subaerial drainage that have persisted in large part to the present day. In Suffolk the uncomplicated drainage pattern points to recession to the northwest, with meltwaters escaping directly away from beneath the ice to the east and southeast, responding to the general inclination of the land and linear depressions on the till surface reflecting subdrift channels and valleys (fig. 9.1). Thus probably all the Suffolk rivers, including the Dove-Waveney and the Blyth-Alde-Chet group, were gradually lengthened headward as the ice withdrew, with subglacial giving way to subaerial incision into the drifts and outwash concentrated along valley lines.

In northeast Norfolk, meltwaters flowed southward as the final stages of construction of the Cromer Ridge proceeded. The highest outwash materials exposed in the eastern cliffs, namely the Gimingham Sands and the Brick Kiln Dale Gravels (Banham 1968) are both affected by diapiric structures, but somewhat younger outwash gravels, not so involved, survive on the crest of the Ridge itself. Two sites here are of especial significance (fig. 7.1).

At the south end of Briton Lane, 2 km southsoutheast of Sheringham, some 20 m of unfolded thick current-bedded sands with seams of coarse flint gravel dip gently south, almost filling a deep gap across the Ridge coincident with a 700 metre 'sag' along the crest of the Ridge. The gravels terminate abruptly on the north side with some slumping and reversal of dip in steep-ravined slopes, but south to Gresham and Aylmerton, they grade gently toward the incised heads of several feeders of the Scarrow Beck tributary of the Bure.

Immediately south of Cromer, other sands and gravels occupy a wider col between Felbrigg and Crossdale Street as an outwash fan which underlies Roughton Heath. Their surface lies at 70 m O.D. in the col some 12-15 m below the general level of the Ridge crest. From their northern margin, which is also back-eroded and fretted by spring-sapped ravines, the surface descends gently south to erosional bluffs overlooking the valley-head of Gunton Beck. The Briton's Lane and Roughton Gravels therefore represent the last stage of meltwater movement over the Ridge from the Sheringham and Cromer areas.

The northern flank of the Ridge (as Reid observed in 1880) has been heavily eroded since its initial construction, possibly by Wolstonian meltwaters (Straw 1965) when this later ice-sheet pressed against the Ridge, and certainly during Devensian times, when periglacial niveo-fluvial and soliflual processes were important in extending and enlarging the many ravines (fig. 7.1). The frequent reduction of spur crests between the ravine systems to circumdenuded knolls confirms a long history of denudation through post-Anglian time. The complicated valley/knoll pattern around Cromer is no doubt partly a response through selective erosion of less-resistant materials to the gross contortions within the North Sea Drift.

Southward of the Ridge the original river Bure (fig. 6.1) not only seems to have integrated a number of meltwater streams but to have established its course toward Lowestoft along the escape route for the ponded meltwaters that formerly inundated east Norfolk. This river system has since suffered two sets of modifications; in the northwest interference by Wolstonian ice, and in the east coastal recession cumulative through interglacials to the present day.

Within the limit of Wolstonian glaciation only the Great Ouse, which possibly failed to return to the Lopham gap after glacial deepening of the Fen basin, may be suspected as an Anglian disruption of a major drainage line. Indeed the Hoxnian pattern retained many easterly-directed streams, and the major drainage alterations south of the Humber pertain to Wolstonian glaciation. Why should this be so, when the Anglian ice was presumably as thick as, and was certainly more extensive than, the Wolstonian? A case was made above for widespread stagnation and wastage of Wolstonian ice, first in marginal zones and then generally, giving residual masses in the vales which interfered with the re-establishment of the former drainage pattern. There is no evidence in east Norfolk and Suffolk for a similar course of Anglian deglaciation. Anglian ice flowed southeast from the Nottinghamshire area successively across the Jurassic and Cretaceous cuestas, in contrast to Wolstonian ice which advanced from the north generally along the Mesozoic strike. During deglaciation the Anglian ice may have remained active instead of decaying within the vales, and with frontal recession toward the northwest, the resultant meltwaters would always have been able to use the easterly-directed pre-Anglian dipslope valleys and cuesta gaps, and modifications to the pre-glacial drainage patterns were avoided.

8 Preglacial denudation

The Anglian ice-sheet entered a landscape that displayed the main relief elements visible today, though the relative relief would have been less. This landscape retains, in a fragmentary series of partly base-levelled surfaces, the imprint of several denudational stages during which differential erosion of the underlying Mesozoic rocks was achieved largely by the gradual though not complete, adjustment of river systems to lithology and structure. Although some of these surfaces may have been cut by the sea, none retain identifiable marine sediments, so that conclusions regarding a sequence of marine transgressions and their extents have always been, and no doubt always will be, controversial. Marine sediments of preglacial Pleistocene and late Tertiary age survive in eastern East Anglia and the history of their deposition will be compared broadly with the denudational stages identified elsewhere in eastern and southeastern England.

Though deferred to the next chapter, evidence for the preglacial evolution of eastern England also derives from analysis of river patterns, for although valley forms change through time as denudation proceeds, valley alignments are remarkably persistent.

Central and northern areas

Comprehensive studies of parts of the south Yorkshire coalfield (Sissons 1954), the Magnesian Limestone cuesta south of the Don (Humphries 1958), the Lincolnshire Wolds (Straw 1961a, 1964), the Yorkshire Wolds (Lewin 1969) and Kesteven (Harrod 1972) have revealed relics of planation surfaces, and several general discussions concerning the extent and age of these surfaces have been published

(Linton 1956, 1964; Straw 1963a, 1970; King 1966; Palmer 1967; de Boer 1974).

Not all valley-side benches, spur 'flats' or summit 'bevels' should however be ascribed to regional planation surfaces. Some, although the result of denudation, are undoubtedly controlled in location and form by bedrock characteristics or have been modified directly by glacial or periglacial processes, while others are the depositional surfaces of glacial and periglacial materials. In what follows discussion refers to erosional features, planed across geological outcrops, for which in eastern England planation by fluvial or marine processes is deemed responsible.

In the Lincolnshire Wolds, five dissected planation surfaces of pre-Wolstonian date have been recognized (Straw 1961a). Although the land form of the chalk cuesta is largely the result of subaerial denudation, three of these surfaces may be of marine origin. The Burnham Surface (55-68 m) is best developed in the northern Wolds where its constituent fragments, some cut across the Middle/Upper Chalk boundary, persist on spurs between dry valleys. The gentle cross-gradient of the Surface and the consistent elevation of its 'back' may be explained by a transgressive marine planation to about 70 m O.D. The Kelstern Surface (115-129 m), a higher feature with similar characteristics, is well preserved in the central and southern Wolds, where it passes onto Lower Cretaceous rocks, and was probably produced during a transgression reaching a maximum height of about 130 m O.D. The highest High Street-Bluestone Surface occupies the scarp crest, and is developed on Middle and Lower Chalk and Lower Cretaceous rocks. A broadly undulating feature between 135-168 m O.D., it does incorporate some very flat bevels, one of which south of Ludford cuts cleanly from Lower Chalk to Carstone across the Red Chalk. These particular bevels may be wave-trimmed (Straw 1964) though most of the Surface has been subaerially fashioned.

North of the Humber, Lewin (1969) identified seven planation surfaces of pre-Wolstonian age in the Yorkshire Wolds. Only the Rowley-Kilham Surface (91-114 m) has features strongly suggestive of marine planation, though certain constituent facets of higher surfaces up to about 200 m O.D. may be wave-cut. Correlation by Lewin and by de Boer (1974) with the Surfaces south of the Humber, apparently made on altitudinal criteria alone, are not wholly satisfactory. Morphologically, the Rowley-Kilham Surface bears strong similarities to the slightly higher Kelstern Surface, and they surely relate to the same transgression. Slight relative downwarping of the Yorkshire Wolds by some 10-15 m following regression could account for the height discrepancy.

In Kesteven Harrod (1972) on the basis of detailed field mapping and analysis of borehole information, claimed the presence of at least seven planation surfaces older than the Wolstonian drift of the area. Four of these, ranging between 85 and 158 m O.D. he considered were wave-trimmed. The best-preserved Harrowby Surface (115-132 m O.D.) is cut across the Lower and Upper Lincolnshire Limestone and is correlated firmly with the Kelstern Surface of the Lincolnshire Wolds. Two less-extensive higher surfaces probably relate to an earlier transgression.

These three studies confirm the presence on some hill areas in eastern England of remnants of erosional terraces that suggest planation by both marine and subaerial processes in relation to different base-levels. Though most of the denudation is subaerial, the common claim for marine inundation to about 125-130 m O.D. has more specific geomorphological and geological significance. Evidence for a deeper more extensive submergence is also given and it is opportune to examine over a wider area the possibility of this latter phenomenon (fig. 8.1).

Rice (1965, 1968) and Clayton (1959) have referred to an East Midlands Erosion Surface developed across a variety of rocks on the higher parts of the Jurassic cuesta southwest of Kesteven below 200 m O.D. Because of its gentle gradients and transection by a series of close-spaced west-east valleys suggestive of an initial pattern of streams little influenced by geological structures, submarine erosion of the Surface has consistently been claimed. Rice (1968) wisely notes, however, that such erosion would be the trimming of a land surface already reduced to low relief by subaerial processes, rather than the abrasion of a single widespread platform.

Northward of the Trent, Swinnerton (1936) proposed the term 'Mansfield Plateau' for a relict surface that truncates Coal Measure and Permian rocks at about 180-195 m O.D., and is crossed by the subparallel valleys of Rainworth Water and the rivers Maun and Meden. In 1937 he explained the plateau as part of a plane of marine denudation that crossed the whole of Lincolnshire into Nottinghamshire. Humphries (1958) examined the extension of this plateau north along the bevelled crest of the Magnesian Limestone and supported a marine origin. Outliers of the plateau occur on Bunter Sandstone south of Mansfield at 160 m O.D. and perhaps on the Keuper Marl at Loath Hill (145 m O.D.) west of Southwell.

North of the Don, Sissons (1954) mapped the Bierley Surface (183-217 m O.D.) across Coal Measure rocks and traced a prominent strandline at its back from northeast of Sheffield to Bradford. Though higher and lower marine surfaces and shorelines were also postulated, it is this 217 m shoreline which coincides with a significant change in land form between the coalfield and the Pennine plateau, and with major contrasts in the drainage pattern.

The coastal tracts of the North York Moors carry remnants of a distinct planation surface at about 160-200 m O.D. West of Scarborough, though most of the interfluves are influenced strongly in

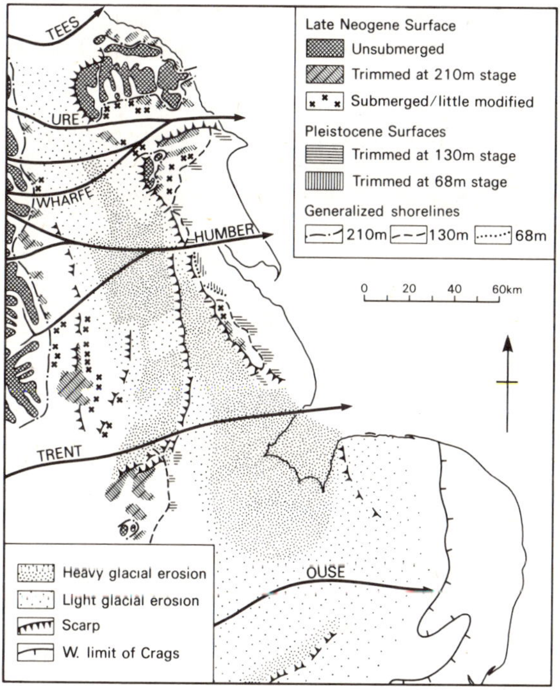

Fig. 8.1 Planation surfaces, zones of glacial erosion and late Tertiary master-streams in eastern England.

their morphology by the underlying Corallian rocks, facets include the higher parts of Irton and Silpho Moors. The southeast part of Harwood Dale Moor lies below 200 m O.D. on Deltaic rocks whereas northward, west of Ravenscar, ground overlooking Robin Hood's Bay, also on Deltaics, reaches over 260 m O.D. On the north side of the Moors a more impressive feature is the shelf of land between Whitby and Saltburn, some three to eight kilometres wide below the 200-metre contour. Though probably a compound feature with erosional facets at different heights, the shelf backs generally against the steeper slopes defining the northern edge of the higher moorland. Separated by several deep valleys excavated by streams frequently heading within these steeper slopes, broad drift-covered interfluves descend generally seaward, those south of Loftus and Roxby carrying much ground between 165 and 200 m O.D. On the west side of the Moors, whereas the Cleveland and Hambleton scarps rise over 300 m O.D., the Howardian Hills nowhere exceed 175 m O.D. In detail the morphology of these Hills is strongly controlled by bedrock lithology and structure, but planation at 160-175 m O.D. on Yearsley Moor in the northwest is closely comparable with the bevels on the scarp crest of the Yorkshire Wolds mapped by Lewin (1969) into his two highest planation surfaces. North of the Vale of Pickering, the crests of two of the Corallian 'flat-irons' of the Tabular Hills between Dovedale and Newtondale drop below 200 m and east of Newtondale a large expanse of Middle Calcareous Grit and Lower Limestone is benched at 175-205 m O.D. around Lockton.

This survey reveals that east of a line from Leeds through Sheffield and Ripley to Loughborough, over a west-east distance of 100 km and excluding only the high plateau of the northwest Yorkshire Wolds, there is no ground higher than 200 m O.D. Ground that approaches this level carries planed surfaces, and features that can be correlated are believed to survive on three sides of the North York Moors, and within the middle Trent vale to the west (Clayton 1953). This accordance of the summits over an area underlain by a wide variety of rocks including several scarp-formers (fig. 1.2) clearly demonstrates regional planation. However, there are some cuesta areas within this zone, for example the Jurassic limestone cuesta northward of the Ancaster gap and the chalk areas of west Norfolk, where preglacial surfaces seem not to have survived, if they once existed. These relatively low cuesta areas were strongly overrun by both Wolstonian and Anglian ice.

Within eastern England, ground higher than 200 m O.D. lies only north of the Humber, and the presence and status of planation surfaces within the Wolds and Moors have been debated frequently since Reed's classic paper in 1901, most recently by de Boer (1974). Lewin (1969) describes the northwest part of the Yorkshire Wolds, reaching to 245 m O.D., as consisting of broadly convex interfluves between deeply incised valleys, which give way east and south at 200 m O.D. to benches of the Wintringham Surface, and west and north to steep scarp slopes.

The higher land of the North York Moors comprises an undulating upland plain that rises from about 245 m O.D. on its east (Fylingdale Moor) and south (Corallian scarp) margins to a culmination at 454 m O.D. on Urra Moors in the northwest. The highest tract coincides with the main Cleveland dome (fig. 1.3), and Versey (1937) considered that the plain had been arched by this uplift and was therefore a feature of Early Tertiary (Palaeogene) date. Versey (1942) extended it to the Yorkshire Wolds, designated it the Wold Peneplane, and identified it also in the eastern Pennines. His view that it was deformed by crustal movement after its formation has been endorsed by de Boer (1974). The plain is however eroded across and is therefore later than both the dome and the Pickering syncline to the south, for whilst it occurs on the Deltaic Moor Grit along the denuded axis, it is on Oxford Clay over Danby Moor north of the river Esk, and on Corallian rocks west of the Dove and north of the Coxwold gap (fig. 1.2). It transgresses the Cleveland Dyke (dated at 58 million years B.P. by Evans *et al.* 1973), and over 300 m of chalk have been removed from the western Yorkshire Wolds (Lewin, 1969). Although in detail many of the constituent facets of this plain now reflect lithological and structural features of the underlying rocks, its regional gradients record not the tilt of a markedly planed surface but the natural fall of the land toward what would have been major drainage axes of the northern part of eastern England, the Tees and the Humber (fig. 8.1). To regard the Wold Peneplane as an unwarped land surface descending from the high ground on the resistant Moor Grit, representing a long period of subaerial denudation through later Tertiary (Neogene) time, is both simple and logical. Such denudation renders pointless a search for traces of a Palaeogene surface on the high ground, but as Lewin notes, the planation surfaces of the Yorkshire Wolds follow rather than create the cuesta backslope, which is an erosional plane warped by the mid-Tertiary Wold syncline. Comparable with the eroded chalk and London Clay surface beneath the older East Anglian Crags and with the lower backslopes of the Chilterns and North Downs, this backslope provides the only feasible location in lowland Britain for a late Palaeogene surface, and any reconstruction of the physiographic history of eastern England before Neogene time exceeds the field evidence and transforms logical argument into speculation.

Field evidence does, however, permit some description also of the form and extent of the Neogene land surface south of the Humber. The compound High Street-Bluestone Surface of the Lincolnshire Wolds, though undulating, incorporates a number of remarkably flat bevels on the scarp crest at 140, 149 and 158 m O.D., which, on the simplest interpretation, were cut at successively higher levels into a land surface during a transgression which eventually submerged the whole of east Lincolnshire (Straw 1964). Rice (1968) made the same point when he suggested that the East Midland Erosion Surface was produced by progressive marine trimming of a landscape already reduced to low relief by prolonged subaerial denudation. Lewin (1969) also concluded that if marine planation was responsible for the Winteringham-Huggate Surface, then the sea flooded a pre-existing landscape.

In a wider context Straw (1963a, 1964, 1970) has argued for a marine submergence of a subaerially-denuded landscape to about 210 m O.D. Summit accordance between the eastern margin of the southern Pennines and the coast is remarkable, but it would be totally unrealistic to postulate one plane surface with inadequate transportational gradients over this vast area. Rather, marine planation of Coal Measure and various Mesozoic rocks at similar heights can be accepted only if the sea is assumed to have drowned a land area of considerable relief. Necessarily a cuesta-and-vale area, some of the clay country west of the chalk must have been flooded before the Lincolnshire chalk was finally submerged. Near the maximum of the submergence deeper water in the drowned vales would permit sufficient wave and current action for the simultaneous trimming of the East Midlands Erosion Surface, the Mansfield Plateau and the Howardian Hills. Marine erosion around the North York Moors would have concentrated on spurs between flooded valleys, and, taking into account planation of the north side of the Moors, the Howardian Hills and the Coal Measures south of Leeds, it seems inevitable that the Moors, like the highest Yorkshire Wolds, became an island at the maximum of the transgression. The submerged land surface was with little doubt the southern and western extension of the end-Neogene plain of the North York Moors, northwestern Yorkshire Wolds and the eastern Pennines (fig. 8.1).

Versey's concept (1937) of a warped surface of minimal relief (the Wold Peneplane) should therefore be replaced by one of a Neogene belted lowland of moderate relief. It is difficult to estimate the height of the vale floors on this lowland relative to present sea-level, but overall gradients along the Wolds crests toward the Humber and on the Jurassic limestones around Ancaster possibly reflect regional slopes prior to 'surface' formation.

This general review of pre-glacial denudation may now, with reference to fig. 8.1, be consolidated into a sequence of postulated events.

The oldest configuration of eastern England that can be detected was a Neogene cuesta and vale landscape, its least-modified part now constituting the upland plain of the North York Moors. Undeformed by crustal movement (except in East Anglia) this land surface was submerged to a height of about 210 m O.D., during a marine transgression which progressed westward along major valleys (Trent, Humber, Pickering and Tees) into the strike vales. In the later stages of the transgression, wave erosion produced benches along the upper backslopes and bevels on the crests of some of the cuestas, to modify the original subaerial slopes. Eventually the Lincolnshire Wolds, Howardian Hills and the Triassic and Permian cuestas were completely submerged. On regression the new wave-cut surfaces, the little-modified slopes of the old land surface, and areas of marine sediment became progressively exposed and subject to attack by subaerial processes. As a regressive sea is not capable of producing extensive abrasion platforms (Bradley 1958), a later and less extensive transgression to 130 m O.D. is considered responsible for development of the Harrowby, Kelstern and Rowley-Kilham Surfaces in Lincolnshire and the Yorkshire Wolds, preceded by regression to at least 115 m O.D. Renewed emergence permitted dissection of these marine platforms, and a period of subaerial base-levelling produced extensive areas of low relief within and flanking the clay vales on a variety of rock types, at 55-85 m O.D. The Burnham Surface in the Lincolnshire Wolds bears witness to a third even more limited inundation to 68 m O.D., and on the analysis of glacial events made above, regression from this level was soon followed by the Anglian Glaciation.

East Anglia

Sparks (1957) identified several erosional terraces within the Cam valley and along the Cretaceous scarp, and claimed their formation to precede the main glaciation of the area. However, no preglacial planation surfaces have been detected on the Norfolk chalk, most probably because of the cumulative effects of glaciation and periglaciation. It has been argued that the Suffolk scarp was pared back by Anglian ice, and that Wolstonian ice here reached its maximum extent. Though Sparks minimized the probable effects of glacial and glacifluvial erosion, the likelihood of such erosion is sufficient to cast doubt on the survival of preglacial forms.

Incontrovertible evidence for marine incursions into East Anglia comes rather from the various deposits of the Crags and the overlying Cromer Forest Bed Series in east Norfolk and Suffolk, all of which, as shown by many coastal sections from Sheringham in the north to Felixstowe in the south, are older than the Anglian Glaciation. It is expedient to review briefly recent work on these deposits in order to ascertain broadly the sequence and date of sea level and shoreline oscillations.

Harmer's (1902) concept of five divisions within the Crags (table 8.1) beneath his Cromerian stage, succeeding one another northward through East Anglia, has never been fully substantiated mainly because of difficulties, when fossil material was sparse or when later reworking had occurred, in allocating local deposits to divisions of the Icenian Crag.

The Crags consist mostly of marine sands and clays which near Great Yarmouth and in the Stradbroke trough are some 70 m thick, thinning west on to the chalk and south on to Tertiary rocks. They lie in the western part of a broad basin which extends across the southern North Sea into the Netherlands. Though overlain in the northeast by the Cromer Forest Bed Series which reaches a maximum thickness of about 10 m, they are largely covered, like the former, by Anglian glacial materials. Recent stratigraphical and environmental information from geological and biological research is clarifying floral and faunal successions and general climatic changes (West 1968; Sparks and West 1972), and there has been some progress in

Table 8.1 Correlation of Lower and Middle Pleistocene stages

Harmer's stages		*Biostratigraphic stages*	*Sediments*
Cromer Forest Bed Series		Cromerian (temperate)	Up to 5m of estuarine silts and sands overlying fresh-water muds and peats and in places a palaeosol.
	I C E	Beestonian (cold)	A few metres of fluvial sands and gravels with an Arctic flora and intraformational ice-wedge casts and involutions.
Weybourne Crag	N I	Pastonian (temperate)	A few metres of estuarine and marine silts and sands.
Chillesford Beds	A N	Baventian (cold)	
		Antian (temperate)	Up to at least 43m of shelly marine sands and marine or estuarine silty clays.
	C R	Thurnian (cold)	
Norwich Crag	A G	Ludhamian (temperate)	
Red Crag		Pre-Ludhamian (cool temperate)	Up to 25m of shallow-water cross-bedded shelly sands and a few thin clay layers
Coralline Crag (Pliocene)		(warm temperate)	5 to 20m of cross-bedded shelly sands

Sources: After Harmer (1902); West and Wilson (1966); West (1968).

relating the new biostratigraphic stages to the classic Crag divisions (Harmer 1902), as indicated in table 8.1.

The Cromer Forest Bed Series outcrops vicariously along the foot of the cliffs of east Norfolk and northeast Suffolk and Cromerian, Beestonian and Pastonian biostratigraphic stages are now recognized within it (West and Wilson 1966). The Cromerian stage has long been accorded the status of an interglacial (Reid 1882; Duigan 1963; West 1968), and the depositional sequence (table 8.1) testifies to a transgressive phase of sea-level movement (West

1972). Cromerian silts are found from West Runton in north Norfolk to Kessingland in Suffolk. Sea level was probably low through the Beestonian when Arctic conditions prevailed, but it is now evident that previously, in Pastonian times, most of the Crag basin was submerged. Near Chillesford and Aldeborough in the southern part of the Crag basin sands and clays of the Icenian Crag ascribed to the Pastonian (West and Norton 1974) rest unconformably on Red and Coralline Crags up to at least 12 m O.D., and this unconformity is traced north into deeper parts of the basin, where older Icenian Crag sediments are preserved. Between Aldeburgh and Lowestoft, the upper Icenian Crag is overlain by the shingly Westleton Beds which Hey (1967) interpreted as the deposits of a beach plain, analogous to the Dungeness foreland, prograded to the southeast. The Westleton Beds, largely overlain by Anglian Lowestoft Till and outwash gravels, contain an abundance of materials transported into the Crag sea by an ancestral Thames. Northward of the Waveney the Pastonian sediments once more consist of estuarine and marine sands, silts and clays, which in the Norwich area rest on subaerially-denuded chalk at about 45 m O.D. (Cambridge 1975). From Chillesford to north Norfolk therefore, the upper Icenian Crag (i.e. Pastonian) including the Westleton Beds, constitutes a series of deposits accumulated during and following the maximum of the transgression.

The older Icenian Crag, known mainly from borehole evidence, consists of up to 43 m of shelly marine sands and silty clays, and comprises four bioclimatic stages (table 8.1). The oldest sediments lie in the deeper parts of the basin, the higher stages onlapping southward until at Chillesford within 15 km of Ipswich, the Pastonian estuarine sediments mark the limit of the transgression.

The Red Crag outcrops within valleys and in coastal sections in an area that encloses the estuaries of the Stour, Orwell and Deben and extends northeast until covered unconformably by the Icenian Crag near Chillesford. Some 25 m of Red Crag occupy the Stradbroke trough in the underlying chalk that trends southwest to northeast between Stowmarket and Bungay (Beck, Funnell and Lord 1972). Elsewhere it is variable in thickness and consists mainly of shallow-water cross-bedded shelly sands including a few thin clay layers. Its surface descends to the northeast and is weathered and oxidized (West and Norton 1974) and its base is strongly unconformable mainly on London Clay but around Orford on the Coralline Crag and in the west on the chalk. In the southern area the transgressive basal layers contain terrestrial vertebrate fossils and reworked Tertiary materials, and are banked against and over cliffs cut in Coralline Crag (Prestwich 1871). The highest Red Crag beach known is at Battisford, 16 km northwest of Ipswich, at 45 m O.D., but most of the Crag is regarded as a regressive deposit (Funnell 1972). Harmer's Red Crag division, based on a molluscan fauna of several hundred species including a high proportion of northern types, is now correlated with the pre-Ludhamian bioclimatic stage (Beck, Funnell and Lord 1972). Formenifera also show an increasing Boreal influence which may reflect either climatic deterioration or severance of the Crag sea from the English Channel and therefore Atlantic influence by relative uplift of areas to the south (Boswell 1952). The Red Crag is now accepted as the oldest British Pleistocene deposit (Boswell 1952; Michell *et al.* 1973), and Beck, Funnell and Lord (1972) suggest a minimum date of 1.6 million years B.P. for the Red Crag in the Stradbroke trough.

The Pliocene Coralline Crag survives only in east Suffolk, mainly within a low ridge at Orford between the Alde and Butley Rivers, but its outcrop, because of a capping of Red Crag and glacial gravels, covers only about 10 km^2. It rests sharply on an irregular surface of London Clay at about 12 m below sea level, but is not found within the Stradbroke and other troughs which, according to Funnel (1972), were probably scoured by strong tidal currents during Coralline Crag times, when the marine connection with the English Channel across southeast England did exist (Wooldridge and Linton 1955). Built into submarine dunes by less-strong currents, the Coralline Crag contains a rich fauna, indicative of warm temperate conditions, but it was eroded and heavily reworked by the Red Crag sea.

It would seem therefore that after deposition of the Coralline Crag, major transgressions into eastern East Anglia took place in lower Red Crag (pre-Ludhamian), late Icenian Crag (Pastonian) and late Cromerian times, separated by periods of net regression.

Such a sequence could reflect either worldwide eustatic shifts of sea-level or local crustal movements or a combination of both. The southern North Sea is known to have subsided through the Pleistocene and the present-day altitudes of the Crags and Forest Bed Series are not the original heights above modern sea-level at which they were deposited. The total amount of subsidence relative to the

Thames basin is of the order of 140 m. West's analysis (1972) indicates that eustatic fluctuations were more important than downwarping during deposition of the Forest Bed Series, and that the pre-Red Crag regression was also eustatically controlled.

Correlation

The high Coralline Crag and Red Crag seas, although separated by a regression of the order of 50-60 m (West 1972) were almost certainly responsible for planations around the lower Thames basin that reached about 200 m O.D., and for deposition of the various Pliocene and Lower Pleistocene sediments (Wooldridge and Lindon 1955). These planations are comparable in form, attitude and height with those in southwest Lincolnshire, around Mansfield, and on the chalk cuestas north of the Fens. There is reason therefore to regard the 210 m transgression over eastern England as contemporaneous with deposition of the Coralline and Red Crags of East Anglia.

The Pastonian transgression reached into Suffolk and on the western shore the Westleton Beds foreland was constructed. Strong lithological and physiographic evidence relates the Westleton Beds to the Pebble Gravels of the lower Thames basin (Wooldridge and Linton 1955; Wooldridge 1960; Hey 1967). In south Hertfordshire the lower Pebble Gravels (Hey 1966) lie between 120 and 135 m O.D. on a dissected platform that declines east at about 1 : 1750. Though generally regarded as fluviatile, Wooldridge (1960) also suggested a marine origin for them. Be this as it may, west of the Braintree 'hinge-line' the Pastonian seems to be represented by the unwarped Pebble Gravels, and an original altitude of at least 120 m O.D. for the maximum of the Pastonian transgression is indicated. This transgression was most likely that during which the Kelstern, Harrowby and Rowley-Kilham surfaces were cut. Cambridge (1975) has provisionally located the Pastonian shoreline in central Norfolk at about 45 m O.D. The highest chalk ground in west Norfolk now barely exceeds 95 m O.D., but must have been considerably higher than 125 m O.D. when the Pastonian shoreline was active. West Norfolk does not appear to lie within the downwarped area, so that the subsequent reduction of the chalk to present altitudes may be a measure of the cumulative effect of glacial and subaerial erosion since then.

Widespread planation of softer rock areas of eastern England at

55-85 m O.D., incorporating a limited transgression to about 68 m O.D. in the Lincolnshire Wolds (Burnham Surface), was described briefly above. By the onset of Anglian glaciation some dissection of this base-levelled surface had occurred, and the Trent had probably cut its valley floor down to about 55 m O.D. in the Ancaster gap. Wooldridge and Linton (1955) describe a similar situation in the lower Thames basin, where an extensive surface about 60 m O.D. (extending beneath the Essex chalky boulder-clay) was dissected to about 45 m O.D. at the Lower Gravel Train stage prior to displacement of the Thames from its course across Essex by Anglian ice. The implication of erosion no lower than about 45 m O.D. in early Anglian times sets a lower limit for the Cromerian transgression in East Anglia. Because no other inundation is recorded in sediments between the Pastonian and Anglian stages, and in Lincolnshire the Burnham Surface is the only one with marine characteristics between the Kelstern Surface and the early Anglian land surface, correlation of the Cromerian and Burnham transgressions seems inevitable.

Comparing the deduced altitudes for the transgressions with the present heights of the deposits in East Anglia, and taking into account relative age and the net eustatic fall of sea-level, there has been some 30 m of subsidence between the maxima of the Red Crag and Pastonian transgressions, about 50 m between those of the Pastonian and Cromerian stages, and about 60 m since Cromerian time. If the Suffolk Red Crag is at least 1.6 million years old then the average subsidence through the Pleistocene has been a little under 1 cm/100 years, a rate which compares closely with that adduced by Sparks and West (1972) for the relative subsidence between Cambridge and the East Anglian coast since the Ipswichian Interglacial.

9 Preglacial drainage development

Several interpretations of the evolution of the river pattern of eastern England have been offered since the latter part of the nineteenth century, the most important being those by Davis (1895), Reed (1901) and Versey (1937, 1942). In essence these authors invoked the agency of river capture along the major soft-rock outcrops during periods of base-levelling, one of which produced a peneplain, now high-level and relict, and another that brought about its dissection while producing the present vales. Windgaps and beheaded streams were evoked to explain the integration of consequent rivers which flowed directly seaward across the Mesozoic rocks. Although Palmer (1967, 1973) proposed an alternative reconstruction, King (1965) and de Boer (1974) reviewed but did not revise the arguments of Reed and Versey concerning the area north of Lincolnshire. However, once glacial modifications are eliminated from consideration (fig. 9.1), it is difficult to point to a single example of such 'Davisian' river capture.

Figure 9.1 illustrates the main features of the pre-Anglian drainage pattern of eastern England and in particular the large proportion of east-flowing rivers and streams. Only the Yorkshire Ouse and the Cam constitute major strike elements and the north-south alignments in the southern part of the North York Moors provide a regional contrast. Most of the rivers west of the Moors and the chalk egress to the sea by one or other of the Tees, Humber, Wash and possibly Lopham gaps.

Drainage and the planation surfaces

In the chalk areas north of the Wash, groups of sub-parallel

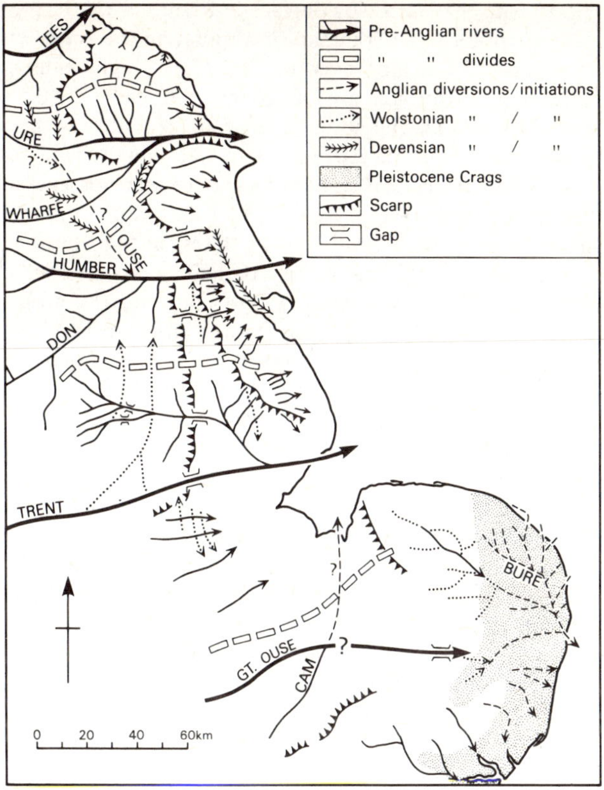

Fig. 9.1 Early Pleistocene drainage pattern of eastern England, with later modifications.

drainage lines are closely associated with both the Burnham and the Kelstern Surfaces in Lincolnshire, and similar lines across the Rowley-Killham Surface in Yorkshire are evident in large sub-parallel dry valleys, especially to the west and north of Driffield. Such patterns, on the assumption that the original streams arose on or were extended across generally plane surfaces, support the interpretation of these surfaces as marine platforms, and valley systems are noticeably more branched and structurally-controlled westward of the 130 m shoreline. Glaciation has obliterated minor drainage associated with the Harrowby Surface in Kesteven, but part of the Jurassic scarp north of Grantham incorporating the Ancaster and Lincoln gaps was planed over at this stage. Significantly, the Humber and Barnetby gaps are located where the chalk crestline also is below 130 m O.D. Initiation of all four extant gaps probably followed regression of the 130 m Kelstern sea, though the history of drainage across the cuestas in both these zones extends back much further (Straw 1964, 1970).

On the Jurassic cuesta south and southwest of the Ancaster gap, the group of east-draining valleys buried by Wolstonian drifts includes the Melton Mowbray-Thurlby valley which headed westward of the scarp (fig. 9.1). Noting the marked parallelism and disregard for structure, Clayton (1959), Rice (1965) and Harrod (1972) have all interpreted the pattern as that of a set of streams initiated on an emergent marine platform, the East Midland Erosion Surface.

North of the Trent, the family of sub-parallel streams in the Mansfield area, together with the River Ryton, its tributary Anston Brook, and the Maltby Dyke further north, are markedly discordant on the Magnesian Limestone and Bunter Sandstone, and until Wolstonian times converged to cross the Keuper Marl by way of the Kirton and Eaton gaps, and the Lias to Lincoln. Northward of Maltby Dyke, the river Don, Hampole Brook, rivers Went, Aire and all other east-Pennine rivers course or once coursed unerringly across the Permian cuesta from Carboniferous country on to Triassic, those north of the Aire having since been glacially deflected.

The epigenetic character of the drainage across the Permian and Triassic rocks argues strongly for regional superimposition at not too distant a point in time. As the areas lie eastward of the general line from Leeds to Nottingham, claimed on morphological evidence as the western limit of marine planation to 210 m, it is not illogical to interpret the sub-parallel rivers as consequent streams that either originated *de novo* on the emergent marine platform or, like the Pennine rivers, were extended across it. Superimposition from younger rocks is also a feasible process, but identification of a cover rock responsible for a single major episode of superimposition along a north-south extent of at least 75 km is not possible. The Trias rocks are themselves crossed by epigenetic streams whilst the dips of the Middle Jurassic rocks and the chalk carry projected westward outcrops well above the requisite levels (fig. 1.2).

The drainage pattern on the flanks of the North York Moors comprises several sets of streams. In the north from Whitby to Guisborough a large group of deeply-incised streams crosses the broad shelf of land below 200 m O.D. to reach the sea near Sandsend, Staithes, Loftus and Saltburn. These streams, their major tributaries and the lower Esk stand in marked contrast to the easterly-directed upper Esk system which, like the upper Don above Sheffield, is contained within a 300 m plateau. The pattern conforms with initiation or re-establishment after a marine inundation to about 210 m O.D.

Two short parallel streams drain northeast into Robin Hood's Bay. Near their heads, Brown Rigg Beck has been glacially converted into a catchwater-type system at the steepish back of a sizeable remnant of the 210 m platform, and it now escapes southeast through higher ground by way of Jugger Howe Beck. The platform remnant from the west side of Stoupe Brow round toward Fylingthorpe betrays a deep embayment in the 210 m shoreline, within which the two streams later originated, perhaps even more pronounced than Robin Hood's Bay within the present coast.

In the Scarborough area the upper Derwent with Lownorth Beck previously drained southeast by Harwood Dale toward Burniston, and the 'Hackness' Derwent reached the sea to the north of Scarborough (fig. 3.3). The latter course is anomalous for the present glacially-induced route of the Derwent 'down' the Corallian dipslope through Forge valley is the more predictable one. Both streams can however be explained as extensions east from the higher moors in accord with a seaward gradient following planation of Irton and Suffield Moors, and once incised into the platform they persisted in these courses until Devensian times. Westward, the northeasterly Corallian scarp of the Howardian Hills is breached in epigenetic manner in the northwest by several streams near Gilling and Hovingham. The latter drain the higher parts of the southwesterly Deltaic cuesta which appears to have been planed over at about 155-175 m O.D.

In summary, there are many drainage elements in the Jurassic areas south of Grantham, along the east Pennine flank and in east and northeast Yorkshire which show such disregard for geological control but association with the highest planation surfaces that they may be claimed in support of the concept of a 210 m inundation of eastern England (figs 8.1 and 9.1). The transgression was controlled by the form of the Neogene land surface, its valleys providing opportunity for the sea to spread inland sooner and further than elsewhere. Sufficient erosion was achieved near the ultimate shoreline and around islands to produce on emergence planed areas on which new groups of sub-parallel streams inevitably arose. Later and more limited marine planations to 130 m and 68 m O.D. account satisfactorily for sub-parallel discordant elements in the drainage systems of the Lincolnshire and east Yorkshire cuestas. Between the inundations, although valleys were deepened and enlarged, very little stream piracy occurred, for no captures can be detected with certainty, and most rivers seem to have maintained their original courses across the Mesozoic rocks until overwhelmed by Anglian ice.

North York Moors

Only the North York Moors preserve a substantial area of land not flooded by the 210 m submergence, and within this area lie both the impressive family of streams draining the south side of the Moors westward from Thornton Dale to the Hambleton Hills and the Esk system to the north of the main divide. Geomorphological interest in the southern streams bears on their close spacing and their transection of the Corallian scarp with a minimal amount of integration along the Oxford Clay outcrop. Their pattern conforms to one of tributaries of a late-Tertiary river flowing east across the Howardians to the sea (fig. 9.1). The middle sections of these streams are being let down gradually on to the Deltaics from the Corallians, and it is logical to assume that the upper sections were superimposed likewise, just as they have since reached into the Lias. This implication that the streams arose entirely on Corallian rocks, which reached at least to the Cleveland axis in early Neogene time, seems more acceptable than the postulation that they were initiated on the southern flank of a warped peneplain developed across the Deltaic, Oxford Clay and Corallian outcrops (Reed 1901; Versey 1937; Palmer 1973), for it avoids the insuperable geomorphological problems relating to perfection of peneplanation, amount of tilt and suppression of strike streams. The close spacing of the streams is primarily a function of the relatively steep gradients reflecting the strong dips off the Cleveland anticline, and retention of separate identity a consequence of the resultant deep incision.

Shedding off the northern flank of the main divide, the right-bank tributaries of the east-flowing Esk constitute a similar but shorter set of streams. Also incised through the Deltaic Series into Lias these streams are gathered by the Esk which has long been accorded the status of a master stream of the region (Reed 1901; Versey 1937; Linton 1951; de Boer 1974). Palmer (1973), however, interpreted the development of the Esk system in terms of initiation of the lower Esk on a marine platform, a strike extension westward along shales near the base of the Upper Deltaic Series, and successive captures of its north-flowing tributaries. This would imply captures of the north-flowing streams relatively late in the Neogene if not in the Pleistocene and abandoned cols across the Cleveland Moors of appropriate depth do not exist. However the Kellaways Rock still occupies much of the Cleveland Moors divide north of the Esk, and piracy by the middle Esk was more probably along the Oxford Clay. Deep incision has since taken the middle Esk into lower strata so that its discordance with structure is the result of 'disadjustment'.

The development of the Esk system clearly does not require a Pennine river, and it seems unlikely that the Esk has ever been a serious competitor with the Tees for Pennine drainage, for the Triassic outcrop must long have reached to the coast.

Mid-Tertiary arching along the west-east Cleveland axis is therefore considered to have produced slopes beneath which the Corallian rocks were largely intact and on which a large number of north- and south-flowing streams were initiated. Breaching of the Corallian cover on the flanks of the arch exposed the Oxford clay, but only on the gentler northern slopes was drainage integration achieved by capture. Uncovering of resistant Deltaic sandstones ensured that the highest ground persisted along the anticlinal axis, so that the regional slopes of the end-Neogene land surface inevitably declined generally toward the Tees, the Vale of Pickering and the east coast. It should be noted that advanced peneplanation and late-Tertiary warping were inferred, not demonstrated, by Reed and Versey solely to explain the physiographic features, and have been endowed since with credibility by repetition in the literature. To claim therefore that these phenomena were responsible for the drainage pattern and regional slopes is a clear case of circular argument.

Neogene master streams

Reed (1901) argued that the east Pennine rivers had formerly maintained their parallel courses to the sea across east Yorkshire, and in particular he carried the upper Tees to the Esk, the Ure to the Rye and thence the Great Wold valley in the Yorkshire Wolds, and the Nidd, Wharfe and Aire/Calder to the Humber. Versey (1942), ignoring the Tees, carried the Swale to the Esk, the Ure perhaps with the Nidd and Wharfe as tributaries to the Rye and thence along the Vale of Pickering, the Wharfe alternatively to the Market Weighton gap, and the Aire/Calder to the Humber. De Boer (1974) combined elements of both these reconstructions, but Palmer (1967) proposed the Ouse instead as the only master consequent.

The concept of the Esk as a beheaded consequent does not stand up to examination, and the Tees is surely the only credible master stream north of the Cleveland axis.

South of this axis, majority opinion, impressed by the Winteringham gap near the head of the Great Wold valley has favoured a course for a former east-flowing river southward of the Vale of Pickering. Palmer (1973) envisaged a simple plane surface across the chalk, Kimmeridge Clay and part of the Corallian to carry the south-flowing streams to it. However, a depression along the Kimmeridge Clay and Speeton Clay outcrops facilitating penetration by the 210 m sea would seem more likely and the Winteringham gap may not in fact merit the status of a major windgap. Its 'floor', some 2 km wide at about 130-135 m O.D., is dissected into benches and low hills by a branched obsequent stream. The extension of this stream and indeed the considerable width of the gap surely relates to the important Foxholes fault which here crosses the scarp. Lewin (1969) found no evidence to support the former use of the gap by Pennine streams, and it is here dismissed as a local feature, the result of rapid scarp recession through the Pleistocene concentrated along the Foxholes fault overtaking a double-headed tributary to the Great Wold valley.

Palmer (1973) in a complex model, claimed capture of the western Tabular Hills streams from the Winteringham gap by a river working from the west at about 100-110 m O.D., followed by a later series of captures from the east. In 1967, impressed by the south-easterly elements in the drainage pattern, he had designated the Yorkshire Ouse the master stream south of the Tees. But why was the western capture not effected through the fault-bounded Coxwold-Gilling gap, and why should a longer western river compete successfully with a shorter Wold river? The more convincing views are those of Reed and Versey, and the interpretation offered here (fig. 9.1) is that through the Neogene, the Ure, Nidd and Wharfe all converged on the western end of the Vale of Pickering, and gathered the Rye and its tributaries to flow seaward along the Speeton and Kimmeridge Clay outcrops north of a low chalk scarp. The height of this clay vale was possibly about 100 m O.D. if the conjoint Nidd-Wharfe is held responsible for planing the south-eastern Howardian Hills at about 110 m O.D. The system was suppressed during the 210 m submergence and it is impossible to say whether this master stream was re-established, or if the Ouse was initiated on regression. Any further speculation is pointless, for by comparison with the situation south of the Humber, it may well have been events during the Anglian or Wolstonian Glaciations that eventually brought the Ouse system into existence.

The Humber represents an old alignment. There are indications that the sea crossed the chalk here early during transgression to the 210 m level (Straw 1970), and the Don, Aire and Calder may all have been headstreams.

The Trent is the only other substantiated Neogene master stream of eastern England (Linton 1951; Straw 1970), but in East Anglia the Lopham depression across the chalk displays the shallowness and breadth envisaged for both the Neogene Humber and Trent valleys. Utilized by Anglian ice, its form may be in part a glacial legacy, but it could have been initiated by a lesser master stream, the Great Ouse, which, with the Cam, drained an area between the catchments of the Trent and the Thames. Whether the Ouse, the Cam or neither passed through it after the 210 m submergence will never be known.

In conclusion, four, or possibly five master rivers crossed eastern England in Neogene times (figs 8.1, 9.1). They and their tributaries were responsible for the production of a cuesta-and-vale landscape, the surface of which ranged from the high ground of the North York Moors approaching 500 m O.D. to the broad valley floors that seem to have been reduced to almost 100 m O.D. before the commencement of Pliocene inundation. That and later inundations inevitably caused modifications to land form and drainage pattern, but the most drastic disruption of the stream systems that had evolved through the Neogene and survived into the Pleisocene occurred during the several glaciations. The drainage pattern of eastern England is therefore composite, compound and complex, but it is a pattern nonetheless, and one that retains a subtle record of its history.

10 The Flandrian interglacial

Introduction

It remains, after the retrospective review of the preceding chapters, to consider the current interglacial period. About 10 300 years have elapsed since the sudden warming of climate that is taken to end the Devensian Glaciation, and within this time, sea-level has risen by some 35 m, mean July temperature has risen by some 6-8° C, and thermophilous flora and fauna have re-entered the landscape.

The precise details of early Flandrian transgression in eastern England, as revealed by sedimentary studies, are not to be related here and reference should be made to Smith (1958), Steers (1964) and Gaunt and Tooley (1974) but it is appropriate to examine briefly features of both the sea-floor and the modern coastline, to discuss the extent and significance of alluviation controlled by the sea-level rise and to consider those fluvial, mass wastage and aeolian processes which have been modified over the later Flandrian by human activity to the extent that no part of eastern England has escaped the hand of Man.

The Flandrian has occasioned such interest that several attempts have been made to recognize and name distinct stages within the period, in terms of either climatic differences or palynologically-determined vegetational change. Within the last two decades, a radiocarbon-based chronology has provided a framework for establishing floral, faunal and climatic change, for regional correlation and for greater precision in the archaeological sequence of industries and cultures. Application of the model interglacial zones proposed by Turner and West (1968) yields an alternative nomenclature. These systems are represented in table 10.1.

Table 10.1 Flandrian zonation, dates and archaeological stages

Zones[1]	*Blytt/ Sernander periods*	*Pollen zones*	*^{14}C years B.P.*	*Archaeological stages*
				Historic
	Sub-atlantic	VIII	1500	
				Roman
Fl III				Iron Age
(late-temperate)			2700	
			3200	Bronze Age
			3600	
	Sub-boreal	VIIb		
				Neolithic
			5200	
			5300	
Fl II (early-temperate)	Atlantic	VIIa		
			7500	Mesolithic
		VI		
	Boreal			
Fl I (pre-temperate)		V	9500	
			9600	
	Pre-boreal	IV		
			10300	
		III		
			10750	
		II		Upper
De IV				
			12000	
		I		Palaeolithic
			14000	

[1] After Turner and West (1968).

The sea-floor

The sea-floor adjacent to eastern England comprises areas of very different character (fig. 10.1). Through the earlier Flandrian the transgressing sea inundated a land surface of some relief, and erosional and depositional processes along the advancing shoreline effected some modification of it. Veenstra (1965, 1969) regards coarser gravels on elevations as lag deposits resulting from selective wave action, and the Devensian materials of Dogger Bank at about

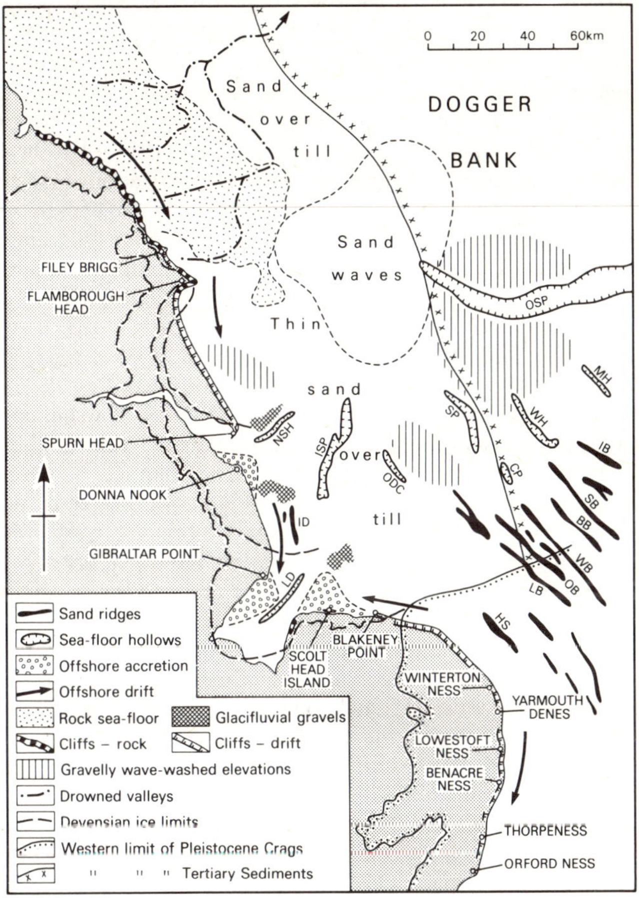

Fig. 10.1 Eastern England — shoreline and sea-floor features. See text for full names of ridges and hollows. (Partly after Dingle, 1965, 1970 and Veenstra, 1969.)

20 m depth as wave-planed. East of Lincolnshire, at depths of 5-15 m, glacifluvial gravels along Devensian moraines have been partly redistributed by wave action (Robinson 1968) and east of East Anglia beach-noded gravels, though probably of Crag provenance, lie in water depths of over 20 m (Veenstra 1969). Dingle (1970) describes a wide offshore outcrop of solid rock east of the North York Moors between about 46 and 70 m depth and suggests that the paucity of tills over it compared with areas south of the latitude of Flamborough Head is a consequence of reworking during the Flandian transgression. This gently-sloping, irregularly-dissected shelf with elevations less than 15 m, gives way eastward to broad hollows which Dingle regards as vestiges of a glacially-modified subaerial drainage system. By contrast, east of Holderness and Lincolnshire, the sea-floor is covered by Devensian tills and is trenched by a number of elongated hollows, including the Inner and Outer Silver Pits, Sole Pit, Outer Dowsing Channel, New Sand Hole, Coal Pit, Well Hole and Markhams Hole (fig. 10.1).

The Inner Silver Pit is about 32 km long, 1.5-2.5 km wide, and reaches up to 60 m below the general sea-floor depth of 20-25 m; New Sand Hole is up to 23 m below it, and Outer Dowsing up to 30 m. Impressed by their collective splayed pattern, Valentin (1957) preferred their subglacial origin beneath Devensian ice, but Donovan (1965) has isolated the Inner Silver Pit as one created by tidal scour, in the manner of The Well channel within the Wash, when sea-level was some 10 m lower than now. Whatever their origin these features are being infilled by sands, rather than eroded, at present.

As modern water depths were approached in the latter half of the Flandrian, wave action became localized along and near to the present coastline, but changes of sea-floor morphology have continued with the movement of sand sediments up to 20 m thick derived essentially from glacial and aeolian materials. Sand sheets are particularly extensive off northeast Norfolk, and over a wide area east of Flamborough Head extending to Dogger Bank (Dingle 1965). In the former area, tidal currents up to 3 knots have moulded the sands into prominent northwest to southeast ridges up to 20 km long such as the Leman, Ower, Well and Swarte Banks (Veenstra 1969). Elsewhere they have built the sands into megaripples that can reach heights of 10 m.

The coastline

The open coast of eastern England from Hartlepool Point south to Landguard Point has a length of 460 km, and has been described in detail by Steers (1964). Geologically it falls into two main sections; north from Flamborough Head it is cut mainly in pre-Tertiary rocks, to the south it is developed across a variety of Quaternary sediments with older rocks exposed only briefly and in widely separated localities. In what is generally a high energy and receding coast, there are exceptional areas of deposition in the Wash and off northwest Norfolk (fig. 10.1). Deposition elsewhere is across estuary mouths and, as with the small East Anglian nesses (fig. 6.1), at other places where complex interactions between offshore and inshore processes occur (Robinson 1966). Much of the eastern East Anglian coast is now protected by concrete sea-walls or wooden barriers. Now rarely reached by waves, cliffs are degrading and vegetating rapidly, but an ominous consequence is the smaller supply of sediment to depositional features. Such an abnormal damping of shoreface dynamics along what remains a high-energy coast augurs ill for these features and paradoxically for the long-term effectiveness of the sea defences. During 'storm surges' (Robinson 1953; King 1972), when water level is raised above the normal tidal oscillation by specific meteorological factors, the whole coastline of eastern England could still be brought under wave attack.

The cliffs between Hartlepool Point and Flamborough Head display much variety of form, especially across the Jurassic rocks of the North York Moors. Lithological and structural diversity within these rocks, which rise in places almost 200 m above sea-level, ensures a striking repetition of inlet, bay and promontory. Wide abrasion platforms occur frequently along this coast, but beaches, which are often bouldery except in Filey Bay, are present only in bays and inlets. Yet the cliffs are rarely wholly vertical and a common element of cliff form is an upper facet comprising a 30-40° slope. Agar (1960) regards those upper facets developed on bedrock as surviving from an older degraded cliff probably of Ipswichian date, though Palmer (1973) expresses different opinions. If Agar's view is correct, then the coastline must approximate in location to that of the Ipswichian Interglacial. These circumstances point not only to a surprising degree of resistance to marine erosion by the Yorkshire Jurassic rocks but also to difficulty in ascribing the wave-cut platforms to later Flandrian erosion. Some preparation of these features during preceding interglacials seems inevitable.

Cliffs cut entirely or partly in glacial drift and affected by landsliding and gullying exist around inlets like Cloughton Wyke, bays such as Robin Hood's and Cayton Bays, and on the Corallian and chalk promontories of Filey Brigg and Flamborough Head respectively. Around Flamborough Head wave attack has exploited joints, faults and bedding planes in the Upper Chalk to produce a wealth of erosional forms, but it has been particularly severe on the north side where, opposite a long northern fetch, the Bempton cliffs rise a sheer 130 m.

The Chalk disappears abruptly at Sewerby where the modern coast transects the Ipswichian coastline (fig. 4.2) and southward the cliffs of Holderness and East Anglia are of very different character in their long and straight or gently curved alignments, continuous if narrow beaches and generally rapid rates of recession (fig. 10.1). The Holderness cliffs, some 58 km long and rising above 30 m O.D. only at Dimlington, reveal excellent if transient sections mainly in Devensian tills. Wasting at a rate of between 1 and 2.75 m per year (Valentin 1971) under both wave attack and incessant slumping, the cliffs yield much sediment, removed southward. By maintaining only narrow beaches, longshore drift facilitates wave attack on the cliffs.

Over the 46 km between Weybourne and Happisburgh the northeast Norfolk cliffs, receding even faster than those of Holderness, display superb sections in Middle Pleistocene glacial and interglacial sediments, and are exposed to wave attack at high tide under strong winds from all but the southwest quarter. Under appropriate conditions eroded materials may travel by longshore drift in either direction along the shore, in part on the beach face and in part in the inshore zone of wave turbulence.

Although severe cliff-foot erosion may locally produce a near-vertical profile, this is usually of short duration, for cliff form is determined primarily by slump, flow and gully processes responding to the lithologically-varied Quaternary sediments. Sustained mainly by groundwater flow toward the cliff face, these processes are much enhanced by wave-spray during storms. Mass movement however is not suppressed in dry periods. Intersecting desiccation cracks form readily along and into the cliff face, rendering large masses of material unstable. Collapse is followed by flow lubricated by entrapped air. Where wave attack is inhibited by sea defences, these processes degrade the cliffs rapidly. Further south the cliffs at Scratby, between Yarmouth and Lowestoft and at numerous localities along the Suffolk coast alternate with stretches of low coast or estuaries (fig. 6.1). Rarely higher than 25 m, the cliffs are frequently steep in response to the very rapid erosion of Crag, Cromerian and Anglian deposits. As these cliffs recede, so do the shingle barriers and spits that depend on them, so that loss of land is prevalent along most of the coast.

In east Norfolk the low coast between Sea Palling and Winterton comprises a mainly sandy beach backed by a narrow belt of dunes. These afford little protection during severe storms to extensive inland marshes now largely reclaimed for agriculture. In 1938, some 3035 ha of marshland were inundated when the dunes at Horsey were breached, and in 1953 the sea broke through the dunes at Sea Palling, though flooding inland was limited (Grove 1953). A concrete sea-wall, commenced after 1938 and extended since 1953 now stiffens the natural defences.

Similar vulnerability characterizes the open Lincolnshire coast, which suffers insidious erosion south from Mablethorpe. Thin sandy beaches, easily eroded during storms, occupy a foreshore developed across late-Flandrian alluvial clays and peats and Devensian till, and are backed by a discontinuous and puny dune belt that has proved insecure during severe storms (Barnes and King 1953). Uncliffed and now protected almost completely by a sea-wall, this coast provides relatively little sediment for longshore drift compared with Holderness and the spit at Gibraltar Point south of Skegness is correspondingly small. Most of the offshore sediment is carried into the Wash embayment (fig. 10.1) on flood tides to aggrade the massive banks northwestward of the eroding ebb channels of Lynn Deeps and The Well (Evans 1965).

The marine sediments and manner of their accretion along the north Norfolk coast west of Weybourne have been studied intensively (Kidson 1961; Steers 1964; Hardy 1964). Although large depositional features such as Blakeney Point and Scolt Head Island confirm the net drift of flint shingle to the west from the cliffs of northeast Norfolk, it must be noted that smaller sand and shingle ridges at Gore Point, Brancaster, Ramsey and Wells Headland are consistent with easterly transportation. Indeed the movement of sediment inshore and offshore is exceedingly complex (Hardy 1964).

Kidson (1961) concludes that offshore material rarely reaches the beach zone, but under a westerly residual tidal stream is carried to the extensive Burnham Flats and adjacent shoals off northwest Norfolk, which do not receive sediment from Holderness (fig. 10.1).

The flint shingle found in such vast quantity along parts of this coast derives mainly from the cliffs east of Sheringham. It is an interesting if incidental point that similar shingle accumulations probably occurred off north Norfolk in previous interglacials. The high concentrations of beach-noded 'cannon-shot' gravel in the Wolstonian drifts of north and central Norfolk may have originated in shingle bars of Hoxnian date, scraped up by Wolstonian ice. The 'eyes', or low hills of gravel in the marshes behind Blakeney Point could be vestiges of an Ipswichian spit. Within the protection of the modern shingle ridges, salt-marshes described in detail by Steers (1960, 1964) have been in part reclaimed for agriculture but elsewhere retain the natural features of tidal creeks and zoned vegetation.

The major estuaries of eastern England, excluding the Wash, display common features in the southerly movement of sediment across their mouths, in the stronger ebb tide currents which ensure incomplete sand and shingle barriers, and in the cyclic evolution of the main spits, the largest of which are Spurn Head and Orford Ness.

Rooted to the southeast corner of Holderness the narrow Spurn Head spit, 5 km long, is oriented northeast to southwest in response to constructive swell waves normal to it, but the sediment comes from the Holderness cliffs. De Boer (1964) has demonstrated how, approximately every 250 years, the spit develops a vulnerability to breaching and subsequent decay concomitant with initiation of a new spit somewhat westward of the earlier feature. The scale of cliff erosion is such that Spurn Head should be a larger feature than it is, but some sediment passes into the Humber estuary on the flood tide, some maintains the offshore Stony Binks bank, and much on the evidence of Robinson's drifter experiments (1964, 1968) travels south across the estuary, to augment in part an obtuse foreland or ness of silt and sand at Donna Nook on south Humberside.

The exceptional and complex shingle spit and cuspate foreland of Orford Ness is some 15 km long (Steers 1964). The northern eroding section, Sudbourne Beach, continues the southerly line of the coast from Dunwich, but the southern Orford Beach faces southeast and is accreting. Longshore drift from the north provides the flints and sand, and Kidson (1963) and Carr (1965) have demonstrated that after periods of growth the distal point of the spit, North Weir Point, is breached during storms. The detached material eventually crosses the estuary to Shingle Street spit, the form of which Kidson (1963) considers to be controlled by counterdrift currents.

Along the open East Anglian coast are five small but remarkable features (figs 6.1, 10.1). Winterton Ness is typical, comprising an obtuse foreland of sand and shingle ridges some 5 km long and 0.5 km broad, covered in part by dunes (Steers and Jensen 1953). The ness moves southward because erosion on the north side is complemented by deposition on the south, and since the early eighteenth century it has, near Winterton, overlapped about 2 km of cliff. Yarmouth Denes, with low dunes, has moved south from Caister within historic times and is now located off the north end of Yarmouth spit. The latter appears to have grown south particularly in Medieval times, and Steers (1964) suggests that it was deliberately shortened to stabilize the mouth of the Yare, that material from the severed distal end eventually augmented the ness at Lowestoft, and that as a consequence the former cliffs between Gorleston and Corton were reactivated. Robinson (1966) argues for tidal channel systems off all the nesses, and it is clear that in their sediment supply, location, form and size the latter, compared with the estuary mouth features, depend more on the local development of these offshore channel systems than on the general configuration of the coast.

In so far as relative location of interglacial shorelines is concerned, a three-fold division of the coast can be proposed. First, the Yorkshire coast north of Sewerby, though probably always a high-cliffed coast, has not receded rapidly and its Flandrian position approximates to its Ipswichian and probably its Hoxnian locations. Second, the present coasts of Holderness, Lincolnshire and north-west Norfolk are not only different in character from, but are well seaward of, the cliffed and regular coasts of earlier interglacials. Third, from Weybourne eastward and southward, the scale of Flandrian recession of the East Anglian coast has been such that conversely the Ipswichian and Hoxnian shorelines probably lay far seaward of the modern coast.

Alluvial areas

The Flandrian rise of sea-level also had a profound effect on the inland physiography of eastern England. Submergence not only brought former land areas under wave attack but also, because rivers were not able to aggrade their valley floors fast enough, eventually carried the sea into the lower reaches of the drainage system converting them into vast estuarine embayments. Latterly however, marine and fluvial deposits have largely infilled these embayments, creating alluvial flats which now comprise some 14% of the land area of eastern England. Lithology and stratigraphy of the sediments is complex, for marine sands, silts and clays interdigitate with peats and freshwater silts and muds. In several areas they have been studied intensively in recent years to establish the stratigraphy, to obtain biological and archaeological data, and to acquire evidence for sea-level movements.

Agar (1954) has elucidated the sedimentary sequence in the estuary and surrounding lowlands of the Tees. Much of the valley floor for 2-3 km on either side of the river below Yarm is overspread by up to 10 m of Late Devensian glacio-lacustrine clays which produce almost level surfaces between 10 and 22 m O.D. Dissection through the early Flandrian later permitted the sea to penetrate to Stockton, and the present estuary is now bordered by extensive spreads of marine sands and silts over 25 m thick. Fluvial alluviation is restricted to the Tees floodplain above Stockton and to the major tributary valleys. Much of the alluvium has been reclaimed as agricultural and industrial building land, in the latter case by the dumping of slag over thousands of hectares, while the estuary has been dredged and partly drained for navigational purposes.

The Vale of Pickering contains no marine alluvium, but clays and silts extend widely over the flat former floor of the Late Devensian proglacial lake, which did not drain completely until well into Fl I. Much of the eastern part of the Vale is floored by peat overlying the lacustrine deposits, but stream alluvium covers only a very small area.

The Humber estuary extends some 62 km inland from Spurn Head to the confluence of the rivers Ouse and Trent, and has a shoreline of 145 km. Marine alluvium less than 10 m thick spreads northwest from Spurn and far up the Hull valley (Valentin 1957; Gaunt and Tooley 1974), but on south Humberside from Grimsby to the chalk it attains thicknesses of 25 m. West of the chalk, the Ancholme valley levels are underlain by silty clays and peats which are at least 8 m thick at Brigg and 27 m at South Ferriby Sluice. Much of the marine clay accumulated during a transgression that terminated during the Iron Age (mid-Fl III). North of the constricting Horkstow moraine the surface was later raised to about 3 m O.D. by post-Roman deposition of a reddish-brown estuarine clay (Smith 1958).

In the Humberhead area, alluvium extends widely about the Ouse below Selby and between the Don and Trent, though it is far less extensive than the Late Devensian '25-foot Drift'. Greyish and brown silts are common materials, the surface layers having been artificially warped (Palmer 1966). The Trent is tidal to Cromwell Lock (80 km upstream) and the Ouse to Naburn Lock (60 km upstream). Although banks to constrain tidal waters were constructed as early as the twelfth century, drainage modifications really began in the seventeenth century with Vermuyden's diversions of the Idle and Don, and levées of the abandoned natural courses of the rivers are still traceable across the alluvial flats as slightly elevated roddons. West of the Isle of Axholme, Hatfield Chase comprises a large raised bog on a foundation of Triassic rocks, that commenced growth in the early 'Atlantic' (Fl II; *c* 6500 B.P.) (Smith 1958). Until the seventeenth century it was separated from Thorne Waste, a similar peaty expanse to the north, by a major branch of the river Don.

Large-scale reclamation of the Humberhead and Ancholme alluvium for agriculture commenced in earnest in the seventeenth century. Warping and inning were essentially late-eighteenth- and nineteenth-century practices, turning to advantage the tidal and sediment characteristics of the Humber. For reasons as yet unknown, the Ancholme valley was not warped. The coastal evolution and history of reclamation of the area of south Humberside between Cleethorpes and Donna Nook have been traced by Robinson (1970). Post-Roman subsidence encouraged the formation of saltmarsh but most accretion has taken place in the past 700 years. Although marine conditions proved particularly favourable to deposition after a stormy Medieval period, it is clear that reclamation and drainage works since 1638 have enhanced it.

The alluvial Outer Marsh of Lincolnshire south of Humberston lies rather below the level of highest spring tides and up to 23 m of sediment rests on an uneven surface of Late Devensian till.

Accretion of saltmarsh clays along what is now an open coast clearly necessitated an offshore barrier to provide the requisite low-energy environment. Swinnerton (1931) suggested a chain of morainic islands but Robinson (1968) had discounted this possibility. Rather, sand and shingle features like those of north Norfolk or perhaps the Netherlands were responsible, developed initially when the water was shallower and the influential Holderness coast was several miles eastward of its present position. Presumably such sandy and shingly barrier islands were destroyed when offshore current and wave conditions changed.

As shown in fig. 10.2 two main types of land occur on the alluvial sediments of the Fenland; the peat areas of the north, west and south and the 'silt-lands' of the northeast which cover almost 56% of the total alluvial area of 3350 km^2 and are continuous in the north with the Outer Marsh. The estuarine silty clays were aggraded to about 3 m O.D. over the seaward tracts of an upper peat during a limited transgression through the late Iron Age. Accumulation of the upper peat had begun in Neolithic time (early Fl III; *c* 4400 B.P.) and through the Bronze Age it built outward and upward from the Fen margin to heights rather more than 3 m O.D., conditions proving particularly favourable in the southern parts of the Fens where meres and raised bogs were interspersed with 'islands' of rock and glacial drift.

The siltlands carry a subtle microrelief, having undergone slight dissection, probably during Roman times, even though tidal silts continued to accrete around the borders of the Wash. From Roman times occupation of the siltlands extended gradually seaward, protected in parts by the so-called Roman Bank. The gain of land from the Wash is an ongoing process and in the past 300 years some 30 400 hectares have been reclaimed, nearly half of this in the seventeenth century. However, much of this reclamation has been made possible only as a consequence of the many engineering works since around 1650. A thousand years ago the shoreline was relatively stable (Kestner 1962) but diversion of the Great Ouse river in the early thirteenth century seems to have caused accelerated accretion west of the new outfall below King's Lynn.

The inland peat areas have been 'improved' by drainage schemes particularly since the seventeenth century (Darby 1956), but their very success has led both to shrinkage of the peat and to its wastage by oxidation, a process hastened by cultivation. The peat surfaces

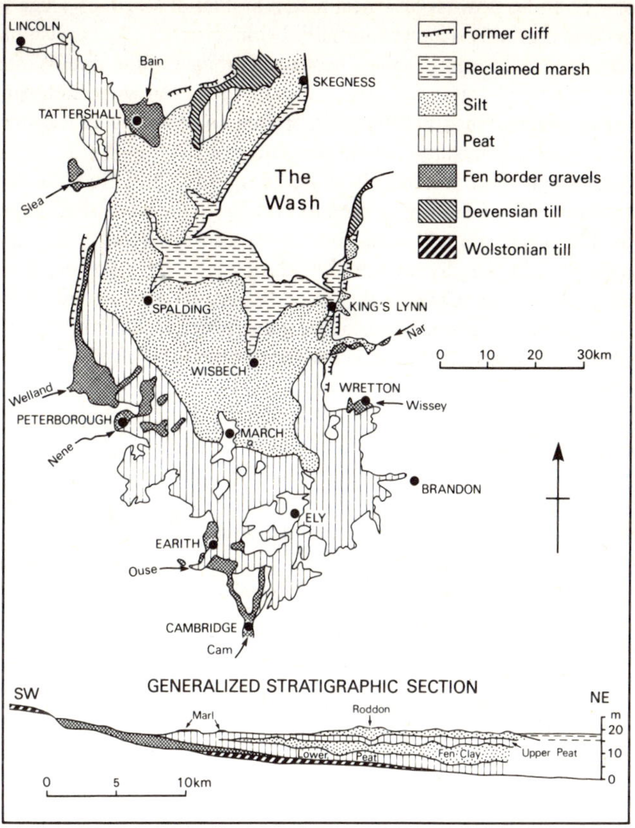

Fig. 10.2 Fen basin and Fen border deposits.

are now up to 5 m lower than those on the silts and wastage has been such that the Fen Clay is exposed over wide areas and linear roddons have become more obvious.

The alluvial flats of east Norfolk and Suffolk lie generally below 3 m O.D. Of broadest extent about the lower Yare and Bure rivers between the 'islands' of Flegg and Lothingland, they finger south-west and north along the valleys of the Waveney, Yare, Bure, Ant and Thurne (fig. 6.1). West of the Yarmouth spit, tidal Bredon Water persists as relic of a former more expansive estuary that reached to

Norwich and is now largely reclaimed marshland. North of Flegg the marshes, probably accreted initially behind barrier spits or islands, lie behind the receding low shore between Sea Palling and Winterton. Drained direct to the sea at Horsey until the early eighteenth century the marshes are still susceptible to flooding as occurred during the storm surge of 1938. Annexed to the Bure and its tributaries are the Norfolk Broads, abandoned and flooded peat-extraction pits of mainly thirteenth and fourteenth century excavation. The origin of the Broads and the details of alluvial stratigraphy have been thoroughly explored and reported by Lambert *et al.* (1960). An Upper Clay marks an early 'sub-Atlantic' (mid Fl III) inundation during which the alluvial surface was built up near to Ordnance Datum and some levée formation took place along the rivers.

The estuaries of southeast Suffolk include those of the Alde/Ore, Deben, Orwell and Stour, but except for the Alde/Ore there has been little examination of their Flandrian sediments. The Deben, Orwell and Stour estuaries all have shorelines about 38 km in length, and the latter two contain vast mudflats but less marsh than that of the Deben. In a study of the Quaternary evolution of the Orford area, Carr and Baker (1968) have described the marsh deposits which they presumed had accreted behind a protective shingle or sand barrier.

Fluvial activity

There has been very little systematic work concerning fluvial action through the Flandrian. Acknowledging valley-floor alluvium to be only a proportion, temporarily resting within the system, of the total material eroded and transported, an examination of alluvial stratigraphy in relation to basin form and characteristics, vegetation development and human occupation would appear to be one way of elucidating the Flandrian history of suitable drainage basins.

With the transition from Devensian to Flandrian climatic conditions, drainage basin response in terms of water and sediment yield can be expected to have changed and even through the Flandrian, environmental change must have ensured variation in water supply to, and its movement within, basins. Independent of climate and vegetation however, denudation can also alter basin response, for example by the progressive exposure of chalk or other permeable rocks with the removal of impermeable drifts. In a number of ways therefore, spatial patterns of runoff in eastern England, controlled by relief, rock type, drift cover and vegetation will certainly have changed through the Flandrian.

Although Clayton (1974) has attempted to place orders of magnitude on the discharge of water and sediment from a group of east coast rivers, present-day fluvial action is imperfectly known, and no catchment studies bearing directly on Flandrian history have been conducted within the area. Rather, interest born of necessity has pertained to observation and documentation of contemporaneous and historical events. One category of such events comprises major river floods. In the Humberhead area, the Yorkshire Ouse and the Trent together with the Derwent, Aire and Don rivers are prone to overtop their banks, particularly along the tidally-influenced reaches. 'Back-up' of freshwater, especially during spring thaws, can produce water levels leading to inundation of alluvial land up to 2 km from the main channels. Nixon (1966) indicates a major flood frequency of about twenty years for the Trent above Nottingham. Radley and Simms (1970) have documented major floods since 1315 on the Yorkshire rivers, and they point as relevant factors to the high discharge of freshwater through the Humber with the convergence of many large rivers, to the extremely low gradients of these rivers, to the presence of much land below high-tide level, and to the insidious subsidence of the estuary area. They also note that, paradoxically, modern engineering works and the progressive raising of flood banks produce conditions leading to a greater frequency of major floods.

Within the Fenlands, the main rivers have in large part been converted to engineered drains designed to carry upland water, as well as Fen water, quickly to the Wash. Drainage schemes in the southeast Fens in the last twenty-five years have incorporated 'catchwater' drains to segregate upland and Fen water (Doran 1965). Although flood situations are commuted where possible by the provision of 'washlands', exceptional spring thaws as in 1947 or back-up of water during storm surges can still inundate much farmland, especially in the peatlands.

Most rivers and streams in eastern England are susceptible to the short-lived 'flash' flood generated by exceptional meteorological conditions. Generally the larger lowland river systems absorb such events with little consequence, but there are many instances concerning streams in the North York Moors and the Yorkshire and Lincolnshire Wolds where very high runoffs have been noted. A typical 'flash' flood was the Bain valley flood of 7 October 1960 (Robinson and Barnes 1961), when 18.4 cm of rain fell within six hours in the Horncastle area. The Fens, the lower Trent and Yorkshire Ouse lowlands have some of the highest thunderstorm frequencies in the country, while the Wolds and the Moors can experience snowfalls from easterly blizzards disproportionately large in relation to their altitude.

Incised courses of streams are restricted mainly to headwater areas. On steeper slopes, particularly in the Moors, they constitute the 'griffs' or ravines which score the head or side slopes of the dales. Some of the ravines have been excavated by surface runoff from adjacent plateau areas (Curtis 1965) others by groundwaters from springs like those at the base of the Corallian rocks, for example in Hacknessdale. Although springs are frequent along the Jurassic and Cretaceous scarps, spring-sapping seems to be a significant contemporary erosional process only along the northwest scarp of the Yorkshire chalk, and within the Bain and Lymn valleys of the southern Lincolnshire Wolds where the Spilsby Sandstone is especially affected. Ravines are notable features of the coastal scenery north of Scarborough, where relatively short, high gradient streams have cut into bedrock through Devensian tills.

The human impact on rivers and streams in eastern England goes beyond the specific schemes of bank reinforcement, channel regulation and water management, to include the more subtle and widespread effects of the substitution by man over the past 5 000 years of a cultural landscape for one dominated by forest. Eastern England in common with much of Britain had acquired, by the middle of Fl I (Godwin 1975), a closed forest which matured over several millenia before clearance was initiated by Neolithic, in places Mesolithic, peoples.

In general under a forest cover, surface runoff is minimized by high infiltration rates and water loss through evapotranspiration, valleyside erosion and supply of coarse material to streams are inhibited, stream regimes are regularized and flood frequency is reduced. Deforestation and cultivation of land are known to exacerbate slope erosion and therefore sediment production within catchments, probably by as much as tenfold. Increase in surface runoff leads to extension of the drainage network and increase in drainage density, for example through gully development. Stream regimes become irregular and with high flood flows more frequent, channel scour where sediment supply is insufficient, and floodplain aggradation where it is excessive, become operative.

When the man-induced changes of the last 5 000 years are eliminated from consideration, an attempt can be made to assess the nature and scale of fluvial activity in the first half of the Flandrian, though in the absence of detailed studies relating alluvial stratigraphy to catchment response, this can only be general.

First, if the sediment supply to and load of rivers are restrained under closed forest, it can be argued that the bulk of the alluvium beneath modern floodplains accumulated before such a vegetation cover was attained. In the early Flandrian this period was also one of rising sea-level which encouraged alluviation. It is probable, therefore, that river floodplains above the estuary zones had largely achieved their present extent by at least 5 000 B.P. if not earlier.

Second, many floodplains are bordered by low terraces or flats of fluvial or glacifluvial gravel, soliflual or colluvial materials or lacustrine sediments. These drifts may be of Late Devensian age, for example the '25-foot Drift' of the Humberhead area and the Floodplain Terrace of the Trent, or Middle Devensian as around the Fens. Slight bluffs continuing beneath the Flandrian alluvium represent the margins of valley incisions that can only, in order to accommodate the alluvium, have been excavated in Late Devensian or earliest Flandrian times, for which period other evidence for low sea-level, steep thalweg gradients and a tundra or birch woodland vegetation exists. Though incision may have commenced considerably earlier outside the limit of Late Devensian ice, within that limit it falls clearly between the clearance of ice from the coastal zone about 15 000 years ago and the return of the sea to the Humber estuary and the Wash by 8 000 B.P.

Third, in most of the hill and upland catchments, a large proportion of slopes retain Late Devensian 'heads' or have been denuded of them only since forest clearance. In East Anglia and Lincolnshire evidence of near-surface patterned ground (fig. 3.8) argues for minimal slope modification. In the North York Moors, it is by contrast with the form and materials of adjacent slopes that ravines are rendered conspicuous rather than by their dominance of the land form. Within the glacierized areas, Late Devensian drifts retain a 'fresh' morphology. Fluvial remodelling of valleyside slopes through the Flandrian has not therefore proceeded very far, and the general conclusion must be that to date basin form, being inherited, has determined contemporary basin processes rather than the reverse.

It can be proposed, therefore, that with reference to fluvial processes, the pre-Neolithic Flandrian comprises two overlapping phases; an earlier one when rates of fluvial erosion and sediment transport were relatively high though declining in intensity, and a later one when, under a developed forest cover and with a higher sea-level, rivers were geomorphologically torpid. Similar progressions characterized the earlier parts of previous interglacials and the interglacial model allows prediction of a resurgence of fluvial activity about the end of the Flandrian, whenever that may be. It can be argued however that during a third (Neolithic and later) phase the tremendous impact of human activity on the landscape has accelerated this predicted end-phase, at least in so far as valleyside erosion and more rapid and irregular runoff is concerned, unfortunately and unprecedentedly at a time of stable high sea-level and minimum gradient on the rivers.

Since the advent of forest clearance and agriculture, rates of soil creep have increased over cultivated land, and mass transfer of soil, even on gentler slopes, is confirmed by accretion on the upslope sides of field walls and hedges. Alluviation of floodplains was enhanced, until constrained by engineering methods, by increases in the suspended load of rivers consequent on Anglo-Saxon and later settlement and cultivation. Within the last few decades, in hilly areas under intensive arable cultivation such as the chalk and Middle Jurassic cuestas, the accelerating transfer of fine materials, producing brashy and less-fertile soils on valleysides and colluvial accretion along foot-slopes and minor valley floors, represents a serious depletion of an important natural resource.

Aeolian activity

The dune-backed coasts of eastern England lie principally in Lincolnshire and north and east Norfolk, where broader beaches provide sufficient quantities of sand for winds from northwest through northeast to southeast quarters to raise a single belt of hills. These dunes rarely exceed 10 m in height, exhibit an undulating crest-line and provide meagre protection against waves during high surge conditions. Only locally, as at Gibraltar Point, Hunstanton, Scolt Head Island and Winterton Ness, does the belt broaden into complexes of low ridges. The dunes are prone to blowout development, often under west and southwest winds, once the vegetation cover has been overtrampled and broken by countless human feet. These same prevailing winds ensure that sand does not penetrate far inland.

Certain inland areas of glacifluvial, glacio-lacustrine, and aeolian sands and gravels suffer wind erosion. The sands along the southern margin of the Vale of Pickering, the Breckland sands and the Cover Sands in the Vale of York and east of the lower Trent are easily mobilized when the surface is dry and unprotected by vegetation, often as a consequence of agricultural and forestry activity (Radley and Simms 1967; Robinson 1968; Wilkinson *et al.* 1969). Near Scunthorpe, several small complexes of blowouts and crescentic dunes, former rabbit warrens, are with the exception of Risby Warren now stabilized and afforested.

Elsewhere light soils, particularly those with large fractions of fine sand, on outcrops of sandstones and sandy limestones, on residual sandy drifts in chalk areas, and on the silts and peats of the Fens, are liable to blow under propitious meteorological and surface conditions. Strong winds in Lincolnshire during the spring of 1968 affected at least 8 100 ha by both removal of topsoil (with seeds and fertilizers) from susceptible fields and its deposition on adjacent land (Robinson 1968).

It is clear that in eastern England contemporary wind erosion, like fluvial incision, is a consequence of human failure to maintain protective vegetal cover over the land surface. Whereas fluvial transfer of sediment from catchments can now be monitored carefully, it is impossible to measure the amounts of material redistributed by wind. However the volume of fine-grained material, especially silt grade, physically redistributed by wind during Flandrian, and even within historic time, must be substantial, and should not be underestimated when interpreting the circumstance of silt-rich horizons in contemporary soils (Catt *et al.* 1971, 1974).

Conclusion

In some respects the Flandrian provides for the geomorphologist a partial analogue for the earlier interglacials, particularly in the early rise of sea-level, the attendant aggradation along valley floors and in estuarine areas and the gradual suppression of surface processes as vegetation cover developed. But in the later Flandrian, man has effected environmental modification on a scale that renders this interglacial very different from those which have gone before. The geomorphological development of eastern England, traceable at least from the Neogene, has in the Flandrian reached a threshold. Never before has a biotic agent had the capability or inclination to control the course of landscape evolution, and the geomorphologist will find the near future as difficult to predict as the distant past to describe.

References

Agar, R. (1954), Glacial and post-glacial geology of Middlesbrough and the Tees estuary. *Proc. Yorks. Geol. Soc.* 29, 237-53.

Agar, R. (1960), Post-glacial erosion of the north Yorkshire coast from the Tees estuary to Ravenscar. *Proc. Yorks. Geol. Soc.* 32, 409-27.

Alabaster, C. and Straw, A. (1976), The Pleistocene context of faunal remains and artefacts discovered at Welton-le-Wold, Lincolnshire. *Proc. Yorks. Geol. Soc.* 41, 75-93.

Baden-Powell, D.F.W. (1948), The Chalky boulder-clays of Norfolk and Suffolk. *Geol. Mag.* 85, 279-96.

Banham, P.H. (1968), A preliminary note on the Pleistocene stratigraphy of northeast Norfolk. *Proc. Geol. Assoc.* 79, 507-12.

Banham, P.H. (1971), Pleistocene beds at Corton, Suffolk. *Geol. Mag.* 108, 281-5.

Banham, P.H. (1975), Glacitectonic structures: a general discussion with particular reference to the contorted drift of Norfolk. In A.E. Wright and F. Moseley (eds), *Ice Ages: Ancient and Modern, Geol. J.* special issue no 6, 69-94.

Banham, P.H. and Ranson, C.E. (1970), A guide to the Pleistocene geology of the Mundesley cliffs. *Trans. Norfolk and Norwich Nat. Soc.* 21, 383-6.

Barnes, F.A. and King, C.A.M. (1953), The Lincolnshire coastline and the 1953 storm flood. *Geogr.* 181, 141-60.

Beck, R.B., Funnell, B.M. and Lord, A.R. (1972), Correlation of Lower Pleistocene Crag at depth in Suffolk. *Geol. Mag.* 109, 137-9.

Bell, F.G. (1970), Late Pleistocene flora from Earith, Huntingdonshire. *Phil. Trans. Roy. Soc. B.* 258, 347-78.

Best, R.H. (1956), Westward proglacial drainage in Cleveland. *Proc. Yorks. Geol. Soc.* 30, 301-19.

Bisat, W.S. (1939a), Older and Newer drift in east Yorkshire. *Proc. Yorks. Geol. Soc.* 24, 137-51.

Bisat, W.S. (1939b), The relationship of the 'Basement Clays' of Dimlington, Bridlington and Filey Bays. *Naturalist*, 133-5, 161-8.

Boswell, P.G.H. (1913), On the age of the Suffolk valleys; with notes on the buried channels of drift. *Q. J. Geol. Soc.* 69, 581-620.

Boswell, P.G.H. (1952), The Pliocene-Pleistocene boundary in the east of England. *Proc. Geol. Assoc.* 63, 301-12.

Boylan, P.J. (1966), The Pleistocene deposits of Kirmington, Lincolnshire. *Mercian Geol.* 1, 339-50.

Boylan, P.J. (1967), The Pleistocene mammalia of the Sewerby-Hessle buried cliff. *Proc. Yorks. Geol. Soc.* 36, 115-25.

Bradley, W.C. (1958), Submarine abrasion and wave-cut platforms. *Bull. Geol. Soc. America*. 69, 967-74.

Bridger, J.F.D. (1976), The Pleistocene succession in the southern part of Charnwood Forest, Leicestershire. *Mercian Geol.* 5, 189-203.

Bristow, C.R. and Cox, F.C. (1973), The Gipping Till: a reappraisal of East Anglian glacial stratigraphy. *J. Geol. Soc. Lond.* 129, 1-37.

Brunstrom, R.G.W. and Walmsley, P.J. (1969), Permian evaporites in North Sea Basin. *Bull. Amer. Assoc. Petrol. Geol.* 53, 870-83.

Cambridge, P. (1975), Field meeting to Bramerton, near Norwich. *Bull. Geol. Soc. Norfolk* 27, 33-46.

Carr, A.P. (1965), Shingle spit and river mouth: short term dynamics. *Trans. Inst. Br. Geogr.* 36, 117-29.

Carr, A.P. and Baker, R.E. (1968), Orford, Suffolk: evidence for the evolution of the area during the Quaternary. *Trans. Inst. Br. Geogr.* 45, 107-123.

Catt, J.A. and Penny, L.C. (1966), The Pleistocene deposits of Holderness, east Yorkshire, *Proc. Yorks. Geol. Soc.* 35, 375-420.

Catt, J.A., Corbett, W.M., Hodge, C.A.H., Madgett, P.A., Tatler, W. and Weir, A.H. (1971), Loess in the soils of north Norfolk. *J. Soil Sci.* 22, 444-52.

Catt, J.A., Weir, A.H. and Madgett, P.A. (1974), The loess of eastern Yorkshire and Lincolnshire. *Proc. Yorks. Geol. Soc. 40*, 23-39.

Clayton, K.M. (1953), The denudation chronology of part of the middle Trent basin. *Trans. Inst. Br. Geogr.* 19, 25-36.

Clayton, K.M. (1959), The geomorphology of the area around Melton Mowbray. *E. Mid. Geogr.* 2, 3-8.

Clayton, K.M. (1974), The glaciated river basins of eastern England. In P. Macar (ed.), *L'evolution Quaternaire des bassins fluviaux de la Mer du Nord méridionale.* Liège, 259-71.

Coope, G.R. (1968), Coleoptera from the 'Arctic Bed' at Barnwell Station, Cambridge. *Geol. Mag.* 105, 482-6.

Cox, F.C. and Nickless, E.F.P. (1972), Some aspects of the glacial history of central Norfolk. *Bull. Geol. Surv. Gt. Brit.* 42, 79-98.

Curtis, L.F. (1965), Soil erosion on Levisham Moor, North York Moors. *Br. Geom. Res. Group Occasional Paper* 2, 19-21.

Darby, H.C. (1956), *The draining of the Fens*, 2nd ed. Cambridge.

Davis, W.M. (1895), The development of certain English rivers. *Geogr. J.* 5, 127-46.

de Boer, G. (1945), A system of glacier lakes in the Yorkshire Wolds. *Proc. Yorks. Geol. Soc.* 25, 223-33.

de Boer, G. (1964), Spurn Head: its history and evolution. *Trans. Inst. Br. Geogr.* 34, 71-89.

de Boer, G. (1974), Physiographic evolution. In D.H. Rayner and J.E. Hemingway (eds) *Geology and mineral resources of Yorkshire*, Yorks. Geol. Soc. 271-91.

de Boer, G., Neale, J.W. and Penny, L.F. (1958), A guide to the geology of the area between Market Weighton and the Humber. *Proc. Yorks. Geol. Soc.* 31, 157-209.

Dimbleby, G.W. (1952), Pleistocene ice wedges in north-east Yorkshire. *J. Soil Sci.* 3, 1-19.

Dingle, R.V. (1965), Sand waves in the North Sea mapped by continuous reflection profiling. *Marine Geol.* 3, 391-400.

Dingle, R.V. (1970), Quaternary sediments and erosional features of the north Yorkshire coast, western North Sea. *Marine Geol.* 9, 17-22.

Dingle, R.V. (1971), A marine geological survey off the north-east coast of England (western North Sea). *Q. J. Geol. Soc.* 127, 303-38.

Donovan, D.T. (1962), Sea-levels of the Last Glaciation. *Bull. Geol. Soc. America* 73, 1297-8.

Donovan, D.T. (1965), Holocene origin of erosion hollows in the North Sea. *Rept. Challenger Soc.* 3, 43-4.

Donovan, D.T. (1973), The geology and origin of the Silver Pit and other closed basins in the North Sea. *Proc. Yorks. Geol. Soc.* 39, 267-93.

Donovan, D.T. and Dingle, R.V. (1965), Geology of part of the southern North Sea. *Nature* 207, 1186-7.

Doran, W.E. (1965), Flood protection and Fen drainage. In J.A. Steers (ed.), *The Cambridge region*. British Association, 41-50.

Duigan, S. (1963), Pollen analyses of the Cromer Forest Bed Series in East Anglia. *Phil. Trans. Roy. Soc.,* B. 246, 149-202.

Edmonds, E.A. and Dinham, C.H. (1965), Geology of the country around Huntingdon and Biggleswade. *Memoirs Geological Survery of Great Britain.*

Edwards, W. (1937), A Pleistocene strand line in the Vale of York. *Proc. Yorks. Geol. Soc.* 23, 103-18.

Edwards, W. (1938), Glacial geology. *Proc. Geol. Assoc.* 49, 333-43.

Edwards, W., Mitchell, G.H. and Whitehead, T.H. (1950), Geology of the district north and east of Leeds. *Memoirs Geological Survey Great Britain.*

Embleton, C. and King, C.A.M. (1968), *Glacial and periglacial geomorphology,* London.

Evans, G. (1965), Intertidal flat sediments and their environments of deposition in the Wash. *Q. J. Geol. Soc.* 121, 209-45.

Evans, P. (1971), Towards a Pleistocene time-scale. Part 2 of *The Phanerozoic Time Scale — a supplement*, spec. pub. Geol. Soc. Lond. no 5, 123-356.

Evans, A.L., Fitch, F.J. and Miller, J.A. (1973), Potassium-argon age determinations on some British Tertiary igneous rocks. *J. Geol. Soc. Lond.* 129, 419-43.

Evans, H. (1975), The two-till problem in west Norfolk. *Bull. Geol. Soc. Norfolk*. 27, 61-75.

Funnell, B.M. (1972), The history of the North Sea. *Bull. Geol. Soc. Norfolk* 21, 2-10.

Gaunt, G.D. (1976), The Devensian maximum ice limit in the Vale of York. *Proc. Yorks. Geol. Soc.* 40, 631-7.

Gaunt, G.D., Bartley, D.D. and Harland, R. (1974), Two interglacial deposits proved in boreholes in the southern part of the Vale of York and their bearing on contemporaneous sea-levels. *Bull. Geol. Surv. Gt. Brit.* 48, 1-23.

Gaunt, G.D., Coope, G.R. and Franks, J.W. (1971), Quaternary deposits at Oxbow opencast coal site in the Aire valley, Yorkshire. *Proc. Yorks. Geol. Soc.* 38, 175-200.

Gaunt, G.D., Coope, G.R., Osborne, P.J. and Franks, J.W. (1972), An interglacial deposit near Austerfield, southern Yorkshire. Report no. 72/4, *Inst. Geol. Sci.* 13pp.

Gaunt, G.D., Jarvis, R.A. and Matthews, B. (1971), The late Weichselian sequence in the Vale of York. *Proc. Yorks. Geol. Soc.* 38, 281-4.

Gaunt, G.D. and Tooley, M.J. (1974), Evidence for Flandrian sea-level changes in the Humber estuary and adjacent areas. *Bull Geol. Surv. Gt. Brit.* 48, 25-41.

Girling, M.A. (1974), Evidence from Lincolnshire of the age and intensity of the mid-Devensian temperate episode. *Nature* 250, 270.

Gladfelter, B.G. (1975), Middle Pleistocene sedimentary sequences in East Anglia. In K.W. Butzer and G. Isaacs (eds), *After the Australopithecines*, The Hague, 225-58.

Godwin, H. (1975), *History of the British flora — a factual basis for phytogeography,* 2nd ed. Cambridge.

Gregory, K.J. (1965), Proglacial Lake Eskdale after sixty years. *Trans. Inst. Br. Geogr.* 36, 149-62.

Grove, A.T. (1953), The sea flood on the coasts of Norfolk and Suffolk. *Geogr.* 181, 164-70.

Hardy, J. (1964), The movement of beach material and wave action near Blakeney Point, Norfolk. *Trans. Inst. Br. Geogr.* 34, 53-69.

Hardy, J. (1966), An ebb-flood channel system and coastal changes near Winterton, Norfolk. *E. Midl. Geogr.* 4, 24-30.

Harmer, F.W. (1902), A sketch of the later Tertiary history of East Anglia. *Proc. Geol. Assoc.* 13, 416-79.

Harmer, F.W. (1910), The Pleistocene period in the eastern counties of England. In *Geology in the Field*, Jub. Vol. Geol. Assoc. 103-23.

Harmer, F.W. (1928), On the distribution of erratics and drift. *Proc. Yorks. Geol. Soc.* 21, 79-150.

Harrod, T.R. (1972), Investigation of major events in the geomorphological evolution of south and central Kesteven. Unpublished Ph.D. thesis, University of Sheffield.

Hemingway, J.E. (1958), The geology of the Whitby area. In G.H.J. Daysh (ed.), *A survey of Whitby and the surrounding area*, Eton, Windsor, 1-47.

Hemingway, J.E. (1974), Jurassic. In D.H. Rayner and J.E. Hemingway (eds) *Geology and mineral resources of Yorkshire*, Yorks. Geol. Soc., 161-223.

Hey, R.W. (1967), The Westleton Beds reconsidered. *Proc. Geol. Assoc.* 76, 403-20.

Hey, R.W. (1967), The Westleton Beds reconsidered. *Proc. Geol. Assoc.* 78, 427-45.

Holmes, S.C.A. (1971), The geological mapper and the employment of his results, as illustrated in some areas of southern England. *Proc. Geol. Assoc.* 82, 161-86.

Horton, A. (1970), The drift sequence and subglacial topography in parts of the Ouse and Nene basins. *Inst. Geol. Sci.* Report to no 70/9.

Humphries, E. (1958), Denudation chronology of the Lower Magnesian Limestone outcrop between Doncaster and Mansfield. Unpublished M.Sc. thesis, University of Sheffield.

Kellaway, G.A. and Taylor, J.H. (1952), Early stages in the physiographic evolution of a portion of the East Midlands. *Q. J. Geol. Soc.* 108, 343-66.

Kendall, P.F. (1902), A system of glacier-lakes in the Cleveland Hills. *Q. J. Geol. Soc.* 58, 471-571.

Kendall, P.F. and Wroot, H.E. (1924), The geology of Yorkshire (2 vols) Vienna.

Kent, P.E. (1955), The Market Weighton structure. *Proc. Yorks. Geol. Soc.* 30, 197-227.

Kent, P.E. (1967), Outline geology of the southern North Sea basin. *Proc. Yorks. Geol. Soc.* 36, 1-22.

Kent, P.E. (1974), Structural history. In D.H. Rayner and J.E. Hemingway (eds), Geology and mineral resources of Yorkshire. *Yorks. Geol. Soc.* 13-28.

Kent, P.E. (1975), The tectonic development of Great Britain and the surrounding seas. In A.W. Woodland (ed.), *Petroleum and the continental shelf of North West Europe, Vol. 1, Geology*. London, 3-28.

Kestner, F.J.T. (1962), The old coastline of the Wash. *Geogr. J.* 128, 457-78.

Kidson, C. (1961), Movement of beach materials on the east coast of England. *E. Midl. Geogr.* 2, 3-16.

Kidson, C. (1963), The growth of sand and shingle spits across estuaries. *Zeit. für Geomorph.* 7, 1-22.

King, C.A.M. (1964), The character of the offshore zone and its relationship to the foreshore near Gibraltar Point, Lincolnshire. *E. Midl. Geogr.* 3, 230-43.

King, C.A.M. (1965), The Scarborough district. *British Landscapes through Maps*, 7, Geographical Association.

King, C.A.M. (1966), Geomorphology. In K.C. Edwards (ed.) *Nottingham and its region*. British Association, 41-59.

King, C.A.M. (1972), *Beaches and coasts.* 2nd ed. London.

Lambert, J.M., Jennings, J.N., Smith, C.T., Green, C. and Hutchinson, J.N. (1960), The making of the Broads: a reconsideration of their origin in the light of new evidence. *Roy. Geogr. Soc. Res. Series* 3.

Lambert, C.A., Pearson, R.G. and Sparks, B.W. (1963), A flora and fauna from Late-Pleistocene deposits at Sidgwick Avenue, Cambridge. *Proc. Linn. Soc.* 174, 13-29.

Lewin, J. (1969), The Yorkshire Wolds: a study in geomorphology. *Occasional Papers in Geography*, no. 11, University of Hull.

Lewis, H.C. (1894), *Glacial geology of Great Britain and Ireland.* London.

Linton, D.L. (1951), Midland drainage: some considerations bearing on its origin. *Adv. Sci.* 7, 449-56.

Linton, D.L. (1954), The landforms of Lincolnshire. *Geogr.* 39, 67-78.

Linton, D.L. (1956), *Geomorphology*. In D.L. Linton (ed.), Sheffield and its region. *British Association 24-43.*

Linton, D.L. (1964), Geomorphology of the southern Pennines and parts of the eastern scarplands. In J.A. Steers (ed.), *Field Studies in the British Isles*. London 138-54.

Madgett, P. (1975), Re-interpretation of Devensian till stratigraphy of eastern England. *Nature* 253, 105-7.

Matthews, B. (1970), Age and origin of aeolian sand in the Vale of York. *Nature* 227, 1234-6.

Melmore, S. (1934), The glacial gravels of the Market Weighton area and related deposits. *Q. J. Geol. Soc.* 90, 141-57.

Melmore, S. (1935), *Glacial geology of Holderness and the Vale of York*. York.

Mitchell, G.F. (1960), The Pleistocene history of the Irish Sea. *Adv. Sci.* 17, 313-25.

Mitchell, G.F., Penny, L.F., Shotton, F.W. and West, R.G. (1973), A correlation of Quaternary deposits in the British Isles.

Geol. Soc. Lond, Special Report no. 4, 99pp.

Morris, J. (1853), On some sections in the Oolitic district of Lincolnshire. *Q. J. Geol. Soc.* 9, 317-44.

Neale, J.W. (1974), Cretaceous. In D.H. Rayner and J.E. Hemingway (eds) Geology and mineral resources of Yorkshire. *Yorks. Geol. Soc. 225-43.*

Nixon, M. (1966), The River Trent. In K.C. Edwards (ed.), *Nottingham and its region*. British Association. 144-156.

Nobles, L.H. and Weertman, J. (1971), Influence of irregularities of the bed of an ice sheet on deposition rate of till. In R.P. Goldthwait (ed.), *Till — a symposium*. Columbus. 117-26.

Norris, G. (1962), Some glacial deposits and their relation to the Hippopotamus-bearing beds at Barrington, Cambridgeshire. *Geol. Mag*. 99, 97-118.

Palmer, J. (1966), Landforms, drainage and settlement in the Vale of York. In S.R. Eyre and G.R.J. Jones (eds) *Geography as human ecology* London. 91-121.

Palmer, J. (1967), Landforms. In M.W. Beresford and G.R.J. Jones (eds) *Leeds and its region*. Leeds, 16-29.

Palmer, J. (1973), Geology and relief. In S.R. Eyre and J. Palmer (eds) *The face of north-east Yorkshire*. London, Part 1, 17-72.

Patterson, T.T. and Fagg, B.E.B. (1940), Studies on the Palaeolithic succession in England. No. 11: Upper Brecklandian Acheul (Elveden). *Proc. Prehist. Soc.* 6, 1-29.

Peel, R.F. and Palmer, J. (1955), The physiography of the Vale of York. *Geogr.* 40, 215-27.

Penny, L.F. (1964), A review of the Last Glaciation in Great Britain. *Proc. Yorks. Geol. Soc.* 34, 387-411.

Penny, L.F. (1974), Quaternary. In D.H. Rayner and J.E. Hemingway (eds) Geology and mineral resources of Yorkshire. *Yorks. Geol. Soc.* 254-64.

Penny, L.F. and Rawson, P.F. (1969), Field meeting in east Yorkshire and north Lincolnshire. *Proc. Geol. Assoc.* 80, 193-216.

Perrin, R.M.S. (1961), Soils. In F. Briers (ed.) *Norwich and its region*, British Association, 44-50.

Perrin, R.M.S., Davies, H. and Fysh, M. (1974), Distribution of late Pleistocene aeolian deposits in eastern and southern England. *Nature*. 248, 320-24.

Pike, K. and Godwin, H. (1953), The interglacial at Clacton-on-Sea,

Essex. *Q. J. Geol. Soc.* 108, 261-72.

Posnansky, M. (1960), The Pleistocene succession in the middle Trent basin. *Proc. Geol. Assoc.* 71, 285-311.

Prentice, J. (1950), Sub-surface geology of the Lincolnshire Fenland. *Trans. Lincs. Nat. Union*, 12, 136-9.

Prestwich, J. (1871), Structure of the Crag-beds of Norfolk and Suffolk, with some observations on their organic remains. *Q. J. Geol. Soc.* 27, 115-46, 325-56, 453-96.

Radley, J. and Simms, C. (1967), Wind erosion in east Yorkshire. *Nature* 216, 20-2.

Radley, J. and Simms, C. (1970), *Yorkshire flooding — some effects on Man and nature*. York, Sessions Book Trust.

Rayner, D.H. and Hemingway, J.E. (eds) (1974), *Geology and mineral resources of Yorkshire*. Yorks. Geol. Soc.

Reed, F.R.C. (1901), The geological history of the rivers of east Yorkshire. London.

Reid, C. (1880), The glacial deposits of Cromer. *Geol. Mag.* 7, 55-66.

Reid, C. (1882), The geology of the country around Cromer. *Memoirs Geological Survey Great Britain*.

Rice, R.J. (1965), The early Pleistocene evolution of north-eastern Leicestershire and parts of adjacent counties. *Trans. Inst. Br. Geogr.* 37, 101-10.

Rice, R.J. (1968), The Quaternary deposits of central Leicestershire. *Phil. Trans. Roy. Soc.* A. 262, 459-509.

Robinson, A.H.W. (1953), The storm surge of 31st January-1st February, 1953. *Geogr.* 181, 134-41.

Robinson, A.H.W. (1964), The inshore waters, sediment supply and coastal changes of part of Lincolnshire. *E. Midl. Geogr.* 22, 307-21.

Robinson, A.H.W. (1966), Residual currents in relation to shoreline evolution of East Anglian coast. *Marine Geol.* 4, 57-84.

Robinson, A.H.W. (1968), The submerged glacial landscape off the Lincolnshire coast. *Trans. Inst. Br. Geogr.* 44, 119-32.

Robinson, D.N. (1968), Soil erosion by wind in Lincolnshire, March, 1968. *E. Midl. Geogr.* 4, 351-62.

Robinson, D.N. (1970), Coastal evolution in north-east Lincolnshire. *E. Midl. Geogr.* 5, 62-70.

Robinson, D.N. and Barnes, F.A. (1961), The Horncastle and Bain valley flood. *E. Midl. Geogr.* 2, 52-5.

Sissions, J.B. (1954), The erosion surfaces and drainage system of south-west Yorkshire. *Proc. Yorks. Geol. Soc.* 29, 305-42.

Smith, A.G. (1958), Post-glacial deposits in south Yorkshire and north Lincolnshire. *New Phytol.* 57, 19-49.

Solomon, J.D. (1931), Palaeolithic and Mesolithic sites at Morston, Norfolk. *Man* 31, 275-8.

Sparks, B.W. (1957), The evolution of the relief of the Cam valley. *Geogr. J.* 123, 188-207.

Sparks, B.W. and West, R.G. (1965), Relief and drift deposits. In J.A. Steers (ed.) *The Cambridge Region*, British Association, 18-40.

Sparks, B.W. and West, R.G. (1968), Interglacial deposits at Wortwell, Norfolk. *Geol. Mag.* 105, 471-81.

Sparks, B.W. and West, R.G. (1970), Late Pleistocene deposits at Wretton, Norfolk. I. Ipswichian interglacial deposits. *Phil. Trans. Roy. Soc. B.* 258, 1-30.

Sparks, B.W. and West, R.G. (1972), *The Ice-age in Britain*. London.

Sparks, B.W., Williams, R.B.G. and Bell, F.G. (1972), Presumed ground-ice depressions in East Anglia. *Proc. Roy. Soc. Lond.* A, 327, 329-43.

Steers, J.A. (1960), Physiography and evolution. In J.A. Steers (ed.) *Scolt Head Island*, 2nd ed. Cambridge, 12-66.

Steers, J.A. (1964), *The coastline of England and Wales* 2nd ed., Cambridge.

Steers, J.A. and Jensen, J.A.P. (1953), Winterton Ness. *Trans. Norfolk and Norwich Nat. Soc.* 17, 259-74.

Stevens, L.A. (1959), The interglacial of the Nar Valley, Norfolk. *Q. J. Geol. Soc.* 115, 291-315.

Straw, A. (1957), Some glacial features of east Lincolnshire. *E. Midl. Geogr.* 1, 41-8.

Straw, A. (1958), The glacial sequence in Lincolnshire. *E. Midl. Geogr.* 2, 29-40.

Straw, A. (1960), The limit of the 'Last' Glaciation in north Norfolk. *Proc. Geol. Assoc.* 71, 379-90.

Straw, A. (1961a), The erosion surfaces of east Lincolnshire. *Proc. Yorks. Geol. Soc.* 33, 149-72.

Straw, A. (1961b), Drifts, meltwater channels and ice-margins in the Lincolnshire Wolds. *Trans. Inst. Br. Geogr.* 29, 115-28.

Straw, A. (1963a), The Quaternary evolution of the lower and middle Trent. *E. Midl. Geogr.* 3, 171-89.

Straw, A. (1963b), Some observations on the Cover Sands of north Lincolnshire. *Trans. Lincs. Nat. Union* 15, 260-9.

Straw, A. (1964), An examination of surface and drainage in the Lincolnshire Wolds, with brief consideration of adjacent areas. Unpublished Ph.D. thesis, University of Sheffield.

Straw, A. (1965), A reassessment of the Chalky boulder clay or Marly Drift of north Norfolk. *Zeit. für Geomorph.* 9, 209-21.

Straw, A. (1966), The development of the middle and lower Bain valley, east Lincolnshire. *Trans. Inst. Br. Geogr.* 40, 145-54.

Straw, A. (1967), The Penultimate or Gipping Glaciation in north Norfolk. *Trans. Norfolk and Norwich Nat. Soc.* 21, 21-4.

Straw, A. (1969), Pleistocene events in Lincolnshire: a survey and revised nomenclature. *Trans. Lincs. Nat. Union.* 17, 85-98.

Straw, A. (1970), Wind-gaps and water-gaps in eastern England. *E. Midl. Geogr.* 5, 97-106.

Straw, A. (1973), The glacial geomorphology of central and north Norfolk. *E. Midl. Geogr.* 5, 333-54.

Straw, A. (1974), The glacial geomorphology of central and north Norfolk: a reply to comments. *E. Midl. Geogr.* 6, 95-8.

Straw, A. (1976), Sediments, fossils and geomorphology: A Lincolnshire situation. In D.A. Davidson and M.L. Shackley (eds) *Geoarchaeology*, London, 317-26.

Straw, A. (1979), The age and geomorphological context of a Norfolk palaeosol. In R.A. Cullingford, D.A. Davidson and J. Lewin (eds) *Timescales in geomorphology*. London.

Suggate, R.P. and West, R.G. (1959), On the extent of the Last Glaciation in eastern England. *Proc. Roy. Soc. B.* 150, 263-83.

Swinnerton, H.H. (1931), The Post-glacial deposits of the Lincolnshire coast. *Q. J. Geol. Soc.* 87, 360-75.

Swinnerton, H.H. (1936), The physical history of east Lincolnshire. *Trans. Lincs. Nat. Union* 9, 91-100.

Swinnerton, H.H. (1937), The problem of the Lincoln gap. *Trans. Lincs. Nat. Union* 9, 145-53.

Swinnerton, H.H. and Kent, P.E. (1976), *The geology of Lincolnshire.* 2nd ed. Lincoln.

Sylvester-Bradley, P.C. and Ford T.D. (eds) (1968), *The geology of the East Midlands.* Leicester.

Taylor, J.H. (1963), Geology of the country around Kettering, Corby and Oundle. *Memoirs. Geological Survey Great Britain.*

Thom, B.G. (1973), The dilemma of high interstadial sea-levels during the Last Glaciation. *Progress in Geogr.* 5, 167-231.

Turner, C. (1970), The Middle Pleistocene deposits at Marks Tey, Essex. *Phil. Trans. Roy. Soc. B.* 257, 373-440.

Turner, C. and West, R.G. (1968), The subdivision and zonation of interglacial periods. *Eiszeit. und Gegen.* 19, 93-101.

Valentin, H. (1957), Glazialmorphologische untersuchungen in ost-england. *Abh. Geogr. Inst. Frei. Univ.* Berlin, 4, 1-84.

Valentin, H. (1971), Land loss at Holderness. In J.A. Steers (ed.) *Applied coastal geomorphology.* London, 116-37.

Veenstra, H.J. (1965), Geology of the Dogger Bank area, North Sea. *Marine Geol.* 3, 245-62.

Veenstra, J.H. (1969), Gravels of the southern North Sea. *Marine Geol.* 7, 449-64.

Versey, H.C. (1937), The Tertiary history of east Yorkshire. *Proc. Yorks. Geol. Soc.* 23, 302-14.

Versey, H.C. (1942), The build of Yorkshire. *Naturalist*, 27-37.

Watt, A.S. (1955), Stone stripes in Breckland. *Geol. Mag.* 92, 173-4.

Watt, A.S., Perrin, R.M.S. and West, R.G. (1966), Patterned ground in Breckland: structure and composition. *J. Ecol.* 54, 239-58.

West, R.G. (1957), Interglacial deposits at Bobbitshole, Ipswich. *Phil. Trans. Roy. Soc. B.* 241, 1-31.

West, R.G. (1956), The Quaternary deposits at Hoxne, Suffolk. *Phil. Trans. Roy. Soc. B.* 239, 265-356.

West, R.G. (1961), The glacial and interglacial deposits of Norfolk. *Trans. Norfolk and Norwich Nat. Soc.* 19, 365-75.

West, R.G. (1968), *Pleistocene geology and biology.* London.

West, R.G. (1972), Relative land/sea-level changes in southeastern England during the Pleistocene. *Phil. Trans. Roy. Soc. A.* 272, 87-98.

West, R.G. and Banham, P.H. (1968), Short field meeting on the north Norfolk coast. *Proc. Geol. Assoc.* 79, 493-512.

West, R.G., Dickson, C.A., Catt, J.A., Weir, A.H. and Sparks, B.W. (1974), Late Pleistocene deposits at Wretton, Norfolk. II. Devensian deposits. *Phil. Trans. Roy. Soc. B.* 267, 337-420.

West, R.G. and Donner, J.J. (1956), The glaciations of East Anglia and the East Midlands: a differentiation based on stone orientation measurements of the tills. *Q. J. Geol. Soc.* 112, 69-91.

West, R.G. and Norton, P.E.P. (1974), The Icenian Crag of south-east Suffolk. *Phil. Trans. Roy. Soc. B.* 269, 1-28.

West, R.G. and Wilson, D.G. (1966), Cromer Forest Bed Series. *Nature* 209, 497-8.

West, R.G. and Wilson, D.G. (1968), Plant remains from the Corton Beds at Lowestoft, Suffolk. *Geol. Mag.* 105, 116-23.

Wilkinson, B., Broughton, W. and Parker-Sutton, J. (1969), Survey of wind erosion on sandy soils in the East Midlands. *Exper. Husbandry* 18, 52-9.

Williams, R.B.G. (1964), Fossil patterned ground in eastern England. *Biul. Peryglac.* 14, 337-49.

Williams, R.B.G. (1968), Some estimates of periglacial erosion in southern and eastern England. *Biul. Peryglac.* 17, 311-35.

Williams, R.B.G. (1969), Permafrost and temperature conditions in England during the last glacial period. In T.L. Péwé (ed.) *The Periglacial Environment* Montreal, 399-410.

Williams, W.W. (1956), An east coast survey — some recent changes in the coast of East Anglia. *Geogr. J.* 122, 317-34.

Woodland, A.W. (1970), The buried tunnel-valleys of East Anglia. *Proc. Yorks. Geol. Soc.* 37, 521-78.

Wooldridge, S.W. (1960), The Pleistocene succession in the London Basin. *Proc. Geol. Assoc.* 71, 113-29.

Wooldridge, S.W. and Linton, D.L. (1955), *Structure, surface and drainage in south-east England.* London.

Worssam, B.C. and Taylor, J.H. (1969), Geology of the country around Cambridge. *Memoirs Geological Survey Great Britain.*

Wyatt, R.J. (1971), New evidence for drift-filled valleys in north-east Leicestershire and south Lincolnshire. *Bull. Geol. Surv. Gt. Brit.* 37, 29-55.

Wyatt, R.J., Horton, A. and Kenna, R.J. (1971), Drift-filled channels on the Leicestershire-Lincolnshire border. *Bull. Geol. Surv. Gt. Brit.* 37, 57-79.

Wymer, J.J. and Straw, A. (1977), Hand-axes from beneath glacial till at Welton-le-Wold, Lincolnshire and the distribution of palaeoliths in Britain. *Proc. Prehist. Soc.,* 43, 355-60.

PART II
The Midlands and Southern Pennines

Keith M. Clayton

11 Stratigraphy and structure

Introduction

The central area of England and Wales covered in this section lies astride the North Sea — Irish Sea water-parting and the Exe — Tees line, the traditional boundary between Upland and Lowland Britain (fig. 11.1). The southern Pennines jut southward towards the Midlands and only 75 km separates Mow Cop from the eastern edge of the Welsh Upland in the Long Mynd. Plains, vales and low scarplands form a U-shaped rim around the southern Pennines: in clockwise sequence the scarplands between the Coal Measure outcrop and the Trent, the Jurassic scarplands of Leicestershire and Northamptonshire, the Midlands, and the Cheshire Plain. While the Midlands are predominantly lowland, several of the horsts that diversify the area rise above 200 m, and almost reach or even exceed 300 m in Charnwood Forest, the Clent Hills, and the Wrekin. Although south of our area the highest summits of the Cotswolds and of the Wiltshire chalk reach similar elevations, summits on the scarplands east and southeast of the Pennines rarely exceed 200 m.

The Palaeozoic/Permo-Trias unconformity

The unconformity below the Permo-Triassic (New Red) rocks of the English Midlands is a critical element in the geography of this area (Kent 1949). Where that unconformity is least well-defined, east of the Pennines where the Magnesian limestone lies with only a small angular disconformity on the Coal Measures, the upland-lowland break is least clear. But in other areas this boundary lies close to a significant landscape break, as along the steep western side of the Pennines, round the Midland coalfield horsts, or in the striking

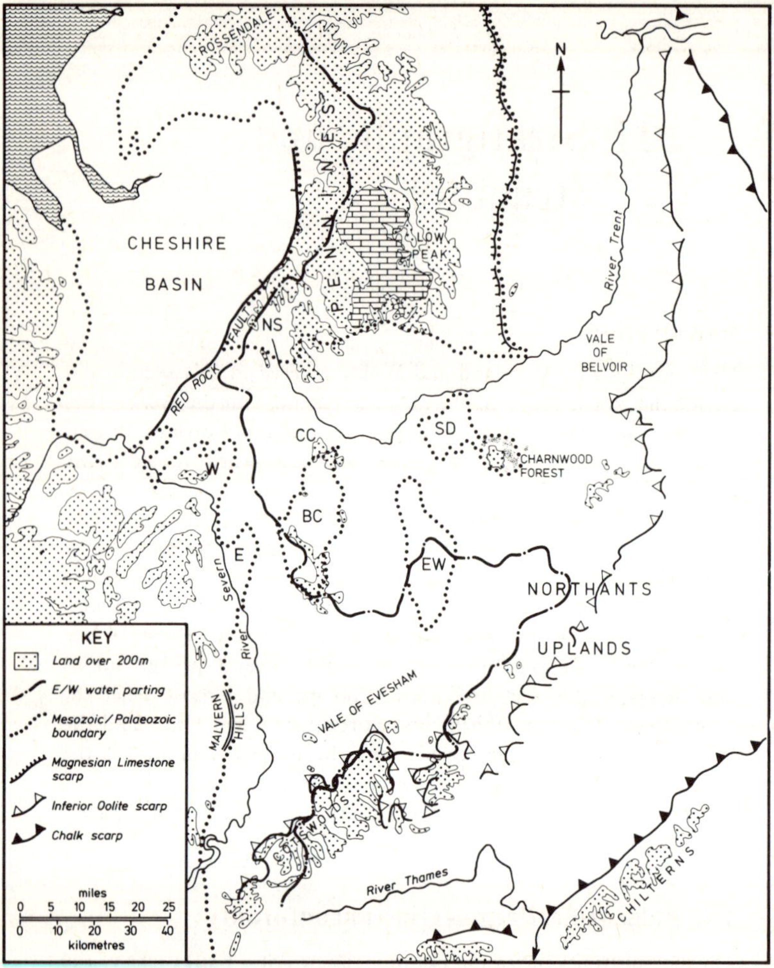

Fig. 11.1 The main features of the relief of the Midlands.

Key to letters:

BC	Black Country	EW	East Warwickshire coalfield
CC	Cannock Chase	NS	North Staffordshire coalfield
E	Enville horst	SD	South Derbyshire coalfield
		W	Wrekin

exhumed landscape of Charnwood Forest. The contrast between the Palaeozoic and Mesozoic rocks is considerable. The older rocks are generally resistant and are often quite steeply folded. In contrast the lower part of the Mesozoic strata contains the thick unresistant beds of the Keuper Marl and the Lower Lias Clay. Above them relatively thin beds such as the Jurassic limestones of the Cotswolds and the Permian Magnesian limestone east of the Pennines form cuestas. The contrast in the angle of dip is also considerable, for since the end of the Carboniferous very little folding has affected the Midlands; the Mesozoic rocks rarely show a dip steeper than a degree or two.

The surface below the Permo-Triassic rocks is, then, one of the more significant unconformities in the British Isles. As a result of the search for coal, and later for oil, its form is quite well known (fig. 11.2). Between the coalfield horsts of the Midlands and the Southern Pennines is a broad basin that opens out eastward beyond Charnwood Forest. To the west it is closed by the rise of the Carboniferous rocks in the north Staffordshire coalfield, itself abruptly terminated on the west by the Red Rock fault which forms the eastern boundary of the very deep Cheshire Basin. Around the margins of the various Palaeozoic inliers more detail can be seen, and in these areas the complex shape of the sub-Permo-Triassic surface shows the extent to which these basin contours suppress these minor, yet significant, irregularities. As Watts (1903, 1947) showed in the Charnwood Forest area, below the surface is an irregular and rather arid landscape that was gradually buried in the sands, pebble beds and clays of the Permian and Triassic periods. Where the unconformity slopes steeply it is likely that this is original, for there is nothing in the folding of the younger rocks to suggest that it is the result of earth movements. This relationship is important because the island-like appearance of the coalfield uplands of the Midlands and the larger upland of the south Pennines has often led geologists to suggest that these have been exhumed in unmodified form by stripping of the Triassic cover rocks. The belief that the uplands of much of the British Isles are exhumed from beneath a Triassic cover has frequently been discussed, and we need not waste more words on it here. Fortunately recent dicoveries have shown that many of the outliers of Triassic material are closely associated with deposits of very late Tertiary age (Boulter *et al* 1971), so that the view so often argued by geomorphologists on general grounds now has the support of correlated deposits (Walsh *et al* 1972). We can confidently say that exhumed remnants of the sub-Triassic landscape will be restricted

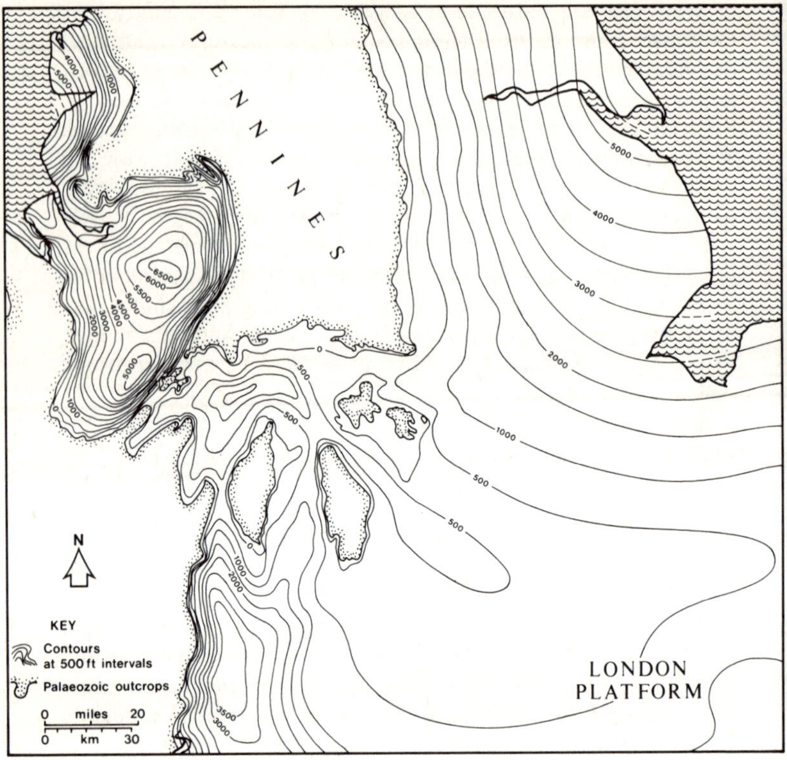

Fig. 11.2 Contours on the Sub-Triassic surface.
Based on the map by P. E. Kent (1949).

to limited areas very near the present outcrop and that they will often slope quite appreciably. Any extensive 'summit plains' cut across Palaeozoic rocks will be of Tertiary or early Quaternary age.

Notwithstanding this, there exists a remarkably good correlation between the Palaeozoic outcrops and upland areas. Here again there has been a temptation, given free play in the absence of any deposits that can establish the age of faulting, to suggest that these uplands owe part of their elevation to movement along the bounding faults in Tertiary times, perhaps during the Alpine (late Oligocene) earth movements of southern Britain (King 1977). Some authors have pointed to the radial arrangement of streams around these as evidence of a general dome-like uplift of a former Triassic cover over each horst. However, it is this feature of the drainage that provides the best evidence that the relief of these Palaeozoic uplands is simply the result of differential erosion. Had there been movement along the faults after the establishment of the main drainage elements of the area, we might have expected these either to maintain their courses across the rising blocks or to be ponded and diverted, giving rise to features that could be detected today. As it is, the very close interrelationship between the Palaeozoic outcrops, outward flowing (i.e. crudely radial) drainage and higher relief, suggests a causal relationship that could only result from the gradual adjustment of drainage to structure over an appreciable period of time and through several epicycles of landscape evolution.

If this argument applies to the Midlands horsts, it is also likely to apply to the southern Pennines. Here again the coincidence of the higher ground and the outcrop of the Carboniferous limestone and more particularly the thick, coarse sandstones of the Millstone Grit series suggests an adjustment to structure that is at least as old as the 600 m summit level of the Kinderscout and Bleaklow plateaux (Linton 1956). The detailed correlation of structure and surface will be discussed in a later paragraph, but there seems no evidence for attributing the relief entirely to late Tertiary movements along the fault lines or fold axes, although some late Tertiary movement cannot be ruled out. Earlier movements of Tertiary age have occurred, but these predated the formation of the highest (600 m) upland surface and, from the point of view of the geomorphologist, are virtually inseparable from the accumulated movement that started in Late Carboniferous times (George 1974). An alternative view has been argued by King (1962) for the Alston Block, and evidence of thick Tertiary basin sequences in the Irish Sea has certainly emphasized the significance of Tertiary tectonics (Dobson 1977a, b), but just how recently these movements have continued is not yet established.

Palaeozoic stratigraphy and structure

Pre-Carboniferous strata

Despite the considerable number of exposures of pre-Carboniferous rocks in the south Midlands it is difficult to generalize about their general outcrop pattern. Along the Welsh borderland exposures are related to the eastern side of the Welsh massif with its broad outcrop of Old Red sandstone and the complex narrow belt of steep folds that brings up the Silurian limestones to form the Abberley Hills ridge north of the Malverns. Here, too, are scattered hills such as Brown Clee and Clee Hill capped by more resistant Upper Old Red sandstones and Carboniferous strata and such striking exhumed ridges as the Wrekin. Further north steep dips are characteristic of both the Carboniferous limestone and the Millstone Grit as are found west of Oswestry and Wrexham. In the Midlands the various inliers may loosely be classified into those brought up in horst-like fault blocks (e.g. south Staffordshire) and exhumed irregularities in the sub-Triassic surface, as at Charnwood Forest. Many of the eastern inliers are associated with more extensive adjacent outcrops of Carboniferous rocks and hardly need special treatment here; their geological interest has assured them extended discussion in geological publications. Examples are the belt of Silurian rocks along the east side of the Warwickshire coalfield, and the outcrops of Silurian limestone at Dudley and Walsall. The western area with its more continuous outcrop shows two contrasted zones, the broad belt of Old Red sandstone outcrop which tapers northwards to terminate northwest of Bridgnorth, and the far more complex area to the west with Silurian cuestas, and the parallel outcrops of the Cambrian and pre-Cambrian rocks. The Old Red sandstone belt west of the Severn forms the western margin of the Midland area. There is a virtually continuous succession from the Ludlow Bone Bed to the Brownstones at the top of the Lower Old Red sandstone — over 1500 m of marls with thin sandstones and occasional thin limestones. At times the Downton Series has been assigned to the Silurian, but the current view is to regard all this sequence as of Old Red sandstone age (Greig *et al* 1968).

Carboniferous strata

For much of the British Isles a tripartite division of the Carboniferous is applicable, with a massive series of limestones at the base, thick shale beds with occasional persistent bands of arkose grit above, and then a gradual change to finer sandstones within the shales and associated with this the incoming of coal seams. This general division into Carboniferous limestone, Millstone Grit and Coal Measures works well in the Pennines and the major geological divisions coincide with zones of contrasted relief. The only extensive outcrop of the Carboniferous limestone is in the Derbyshire Dome where over 700 m of massive Visean limestone is seen and at least as much again has been proved in boreholes (Ford 1977). At Woo Dale the lowest limestones rest on volcanic rocks of pre-Cambrian age, but at Eyam the borehole entered Ordovician slaty mudstones at a depth of 1803 m (Dunham 1973). The limestones contain layers of chert, and in some areas volcanic tuffs and lavas (locally called toadstones). The considerable mineralization is more probably associated with underlying intrusions (as in the Weardale granite of the Alston Block) rather than this volcanic activity. In places the limestone seems to be uncomfortably overlain by the shales at the base of the Millstone Grit series (locally called the Edale Shales and, in the past, the Limestone Shales), but the relationship is not everywhere clear and has caused considerable controversy. The coarse grits of the Millstone Grit series dominate the sequence by their influence on relief although they are in total rather less thick than the shales between them. Individual grits may be traced for tens of miles although their lithology and relative significance as relief formers changes gradually. The top grit of the series, the Rough Rock, can be identified over the whole of the southern Pennines. There is a tendency for the Coal Measures to become rather less resistant up the succession as the sandstones become less important, while both west and east of the Pennines the uppermost part of the Coal Measures is cut out by overlapping Permian or Triassic rocks.

In the Midlands, the Carboniferous succession is very different. The Carboniferous limestone is often absent, although small inliers occur as at Breedon-on-the-Hill, north of Charnwood. Carboniferous limestone was deposited in the southwest of the Midlands (Forest of Dean to Clee Hill) and may have been deposited further northeast only to be removed by erosion. The upper part of the Carboniferous limestone of Derbyshire is generally represented by sandstones or shales along the Welsh border, but eastwards limestones were probably deposited, only to be removed by erosion at a later stage: they are known from pebbles in the youngest Carboniferous conglomerates. The Millstone Grit period (Namurian) is very poorly represented in the Midlands and the initiation of sedimentation in Coal Measure times (Westphalian) took place at different times so that the Upper Carboniferous successions are very varied with 300 m of Lower Westphalian at Cannock and under 150 m at

Coventry. Sedimentation was accompanied by earth movements so that local variations such as conglomerate beds occur: these changes tended to become more sudden towards the end of the Carboniferous period and the uppermost strata of the Etruria Marl, Halesowen Group and Keele Beds are red and lithologically indistinguishable from the New Red sandstone beds of the Permo-Triassic age (Wills 1950).

Some general idea of the structure of the Carboniferous rocks of the Pennines can be obtained from fig. 12.3 which shows structure contours on the Rough Rock, a persistent arkose band towards the top of the Millstone Grit series. The map is based on careful contouring at 1:63 360 but not quite all the area is covered by modern geological maps, while variations in thickness within the Carboniferous rocks make the contour values rather approximate as one passes higher or lower in the geological succession. They are perhaps least reliable over the Carboniferous limestone outcrop in the south since there is something of an unconformity at the top of the limestone and the overlying limestone shales vary a great deal in thickness.

Despite these cautions, the map shows a well-known and yet inadequately appreciated contrast between the western and eastern side of the Pennines. On the east the strata dip at between 70 m in 1 km as just east of the Erewash valley, northwest of Nottingham and again between Sheffield and Rotherham, and the rather widespread value of about 50 m per kilometre which is particularly characteristic of the area south of Leeds. These dips carry right up to the rather narrow (i.e. about 5 km wide) crest of the main Pennine structure, giving the broad dip slopes and impressive westward facing scarps that are so characteristic of these eastern moors. The Coal Measures are overlain by the Magnesian Limestone, along a north-south line as far as the Don and then about 20° west of north, baring an outcrop of Coal Measures that increases steadily in width from 12 km northwest of Nottingham to cover 30 km south of Leeds. The Coal Measures continue to dip gently eastward below the Permian and Mesozoic rocks so that an extensive concealed coalfield is worked almost as far east as the river Trent. The Coal Measures terminate along an east-west line passing 6 km north of Leeds, and as the structure contours show, the structures in this area are influenced by the line of the Craven faults and the southern margin of the Askrigg Block.

In contrast the western side of the Pennines is much more complex with an almost unbroken chain of pitching anticlines and related synclines arranged *en echelon* from the Potteries Coalfield right through to the anticline east of Nelson at Widdup. Most of these folds can only be traced as strong features for 12-15 km, although their central sections show a relative relief of 600-900 m, and even more in some cases. Indeed, from Biddulph Moor towards Stoke-on-Trent the structure contours show a drop of over 2000 m in about 7 km, while southwest of Skipton the rocks rise through 1500 m in only 3 km. Most of these folds lie west of the highest belt of the Pennine structure, and they are bounded to the west by what is generally a faulted junction with the Triassic basin of the Cheshire Plain. This is best developed in the south where the Red Rock fault cuts off the Carboniferous structures of the Potteries Coalfield with a throw of well over 1500 m. North of Manchester the Triassic basin is closed by the westward projection of Rossendale Forest where rather strongly folded Carboniferous rocks lie rather lower structurally than the Pennine Crest to the east, but still high enough to have the resistant Lower Coal Measure sandstones and the Millstone Grit give moorland at similar altitudes to the Pennine Crest. As with the transition in the latitude of Leeds, this marks both a structural adjustment to the rather different pattern of the northern Pennines, and also a transition to the very different sedimentary sequence of the Trough of Bowland south of the Craven faults (Earp *et al* 1961).

Locally, folds not unlike those of the western Pennines occur within the eastern area, as at Ashover and Crich east of Matlock. Others, more deeply buried, trapped the oil at Eakring and other localities northwest of Newark-on-Trent. Few of these folds are fully inverted, that at Ashover being the most striking example where even the core of Carboniferous Limestone (and volcanic toadstone) has been neatly bisected by the river valley (Sweeting 1946). Indeed most of the anticlines have a high uninverted section (e.g. Crich) and only part of the rim has been inverted to an anticlinal vale. Similarly, the synclines may be followed by axial streams, the best example being the Goyt in the western Pennines. Considering the great lithological contrasts within these rocks, this limited adjustment to structure is rather unexpected. The processes involved have been subject to detailed investigation in some areas, with some controversial ideas put forward by Pitty (1965).

Mesozoic stratigraphy and structure

Across the Midlands and the Cheshire Basin the lower part of the Mesozoic succession (here taken to include the Permian) was deposited under desert conditions so that stratigraphical ages have tended to be assigned on a lithological basis and correlations are poorly established (Wills 1970). Much of this problem is quite irrelevant to the geomorphologist and he may be content with the commonly-used lithological subdivision into Bunter pebble beds and dune sands and the Keuper Series, the greatest part of which consists of red marls with some thin, more resistant, beds of sandstone. East of the Pennines the sequence is very different, the Permian Magnesian Limestone oversteps the Coal Measures and the lower part of the Keuper consists of sandstones (the Waterstones). West of the Pennines the sequence is very much like that of the Midlands although glacial deposits are so widespread that it is not widely exposed. The Jurassic transgression (heralded by the Rhaetic) led to much greater uniformity, and the Lower Lias occurs in the southeast Midlands, east of the Pennines, and as a small outlier in the Cheshire Basin. There is general uniformity in the Jurassic sequence, although there are variations along the strike in the thickness and lithology of the scarp-formers, while there is a tendency for these beds to thin down-dip.

The Mesozoic rocks occupy a series of deep basins west of the Pennines and to some extent these persist south of the Pennines and between the Midland horsts. To the east and southeast the Palaeozoic floor is much more level: towards London it forms a platform within about 300 m of the surface, in Lincolnshire it slopes steadily eastwards at about 30 m per km (Kent 1949). As we have already noted, in restricted areas such as Charnwood Forest there is clear evidence that the Keuper Marl rests on an extremely irregular land surface that is being exhumed. In other areas within a kilometre or two of the Mesozoic outcrop exhumed patches of a ‘sub-Triassic’ surface may be recognized, but the steep slopes of this surface ensure that it passes high above the present ground.

The Mesozoic sequence is summarized in the table. The occurrence of alternating resistant and unresistant beds means that, where the dip permits, these rocks often form cuesta landscapes (fig. 11.3). This is best seen in the south of the area in the Cotswolds where the dip of 20 m per km, combined with the considerable thickness of the

Inferior Oolite limestone, builds a fine scarp. Below it the Middle Lias marlstone forms a bench, while the higher Great Oolite forms a second plateau level on the dip-slope summit. The higher beds in the Jurassic series are also well-developed in the south, and contribute a series of low scarps, or at least contrasted belts of country. To the northeast they become thinner (and are in part overstepped by the Cretaceous) so the Upper Jurassic there contributes little more than a wide vale developed on the Oxford and Kimmeridge Clays. Even that vale is greatly complicated by glacial drifts and the great alluviated embayment of the Fens.

North of the Cotsworlds true cuesta landscapes are not again found on the Jurassic rocks until the very limit of our region near Grantham. In part this is due to the decline of the Inferior Oolite limestone, the major scarp former both in the Cotswolds and in Lincolnshire. But more important is the decline in the regional dip which in Northamptonshire is often as low as 5-6 m per km. This low angle means that the high cuesta of the Cotsworlds is replaced by a complex plateau area where at first sight little more can be established other than a general correlation between clay-floored valleys and more resistant (sandstone- or limestone-capped) ridges. A small-scale map will show a belted arrangement of the outcrops, but these are very wide and greatly complicated by the areas of older beds exposed in all the main valleys (fig. 12.5). A scarp-like northwestern limit can be identified for the Middle Lias, and this builds a broad dissected plateau that slopes very gently east with the dip. On it occur numerous outliers of the Inferior Oolite (here primarily Northamptonshire Sand) and these increase in size and number down-dip. By the time they have coalesced into a low and reasonably continuous feature that might be called a scarp, we are 16-20 km down-dip. This broad zone of dissection is the combined effect of the low dip and the regional tendency for the resistant beds to thin down-dip so that reduced elevation above base-level is offset by a decrease in resistance to erosion. It is further complicated by the fact that cambering, developed under periglacial conditions, frequently gives valleyward dips well in excess of the regional value (Hollingworth and Taylor 1946).

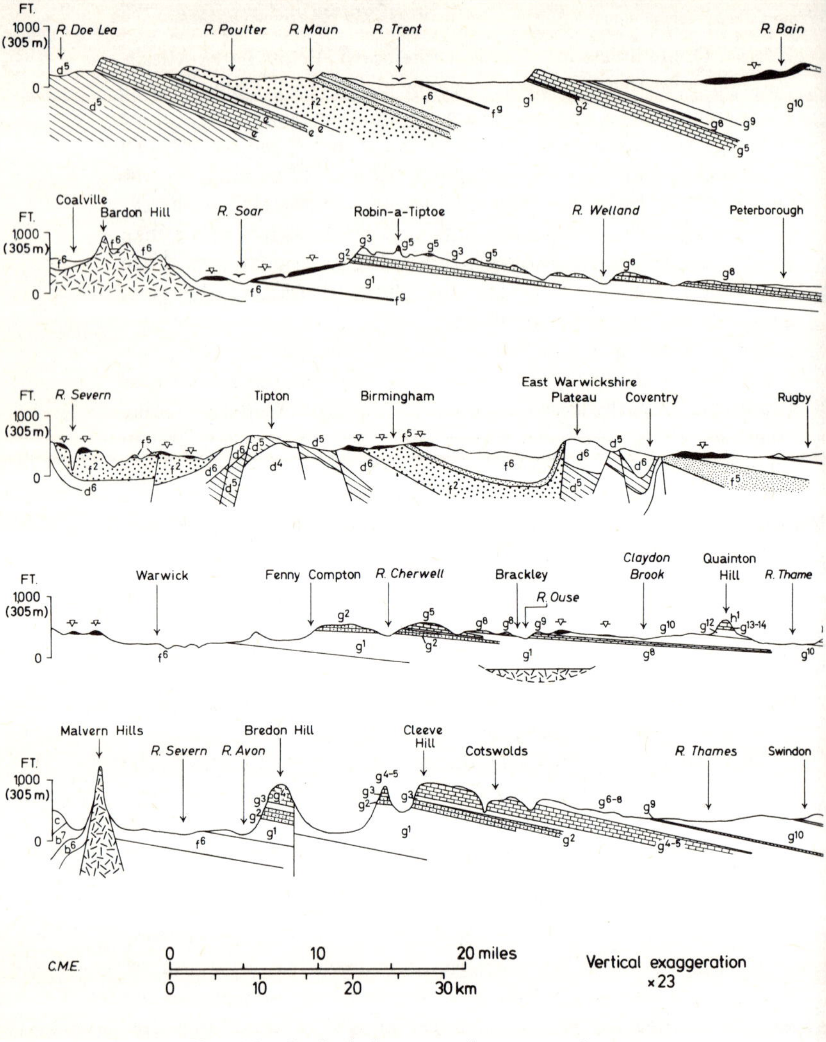

Fig. 11.3 Geological cross sections across the area.

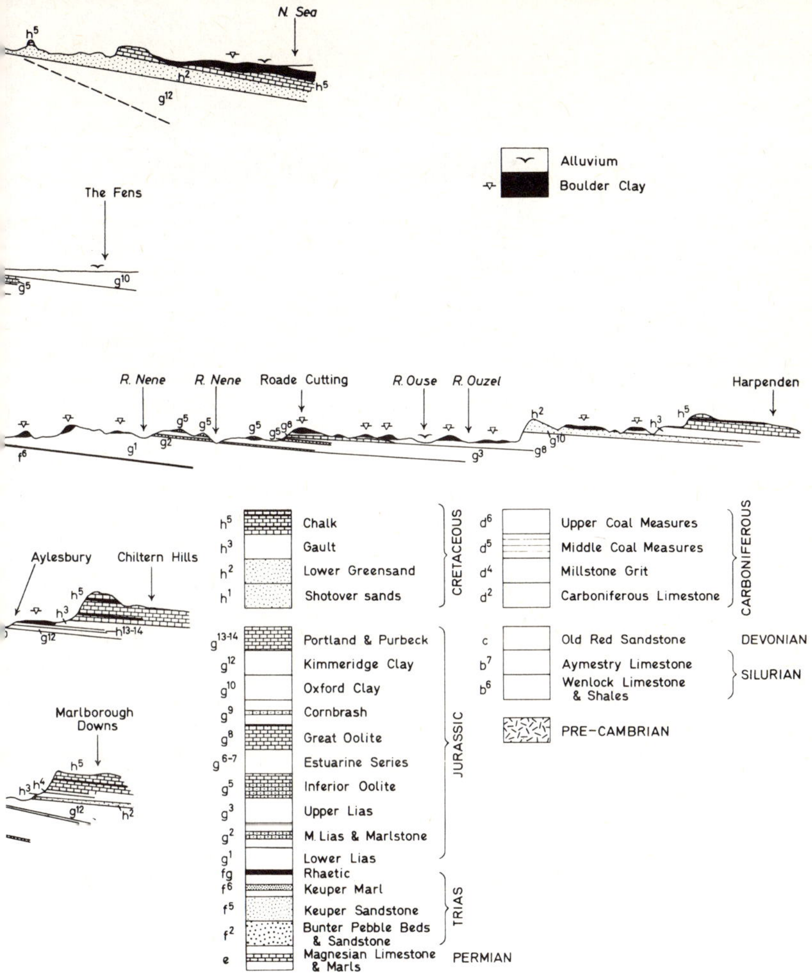

N. Sea
h5
h2
g12
Alluvium
Boulder Clay
The Fens
g5
g10
R. Nene
R. Nene
Roade Cutting
R. Ouse
R. Ouzel
Harpenden
f6
g1
g2
g5
g8
g3
h2
g10
h3
h5
Aylesbury
Chiltern Hills
h13-14
Marlborough Downs
h4
h5
Chalk
Gault
Lower Greensand
Shotover sands
CRETACEOUS
Portland & Purbeck
Kimmeridge Clay
Oxford Clay
Cornbrash
Great Oolite
Estuarine Series
Inferior Oolite
Upper Lias
M. Lias & Marlstone
Lower Lias
JURASSIC
Rhaetic
Keuper Marl
Keuper Sandstone
Bunter Pebble Beds & Sandstone
TRIAS
Magnesian Limestone & Marls
PERMIAN
Upper Coal Measures
Middle Coal Measures
Millstone Grit
Carboniferous Limestone
CARBONIFEROUS
Old Red Sandstone
DEVONIAN
Aymestry Limestone
Wenlock Limestone & Shales
SILURIAN
PRE-CAMBRIAN

12 Lithology and scenery

The Low Peak

The upper part (Visean) of the Carboniferous Limestone outcrops over an area 20 km across and 30 km from north to south in the Derbyshire Dome. Physiographically the outcrop forms a plateau with a summit surface between 350 and 500 m. Surface drainage is restricted to a few deeply-incised rivers, notably the Wye which succeeds in crossing the outcrop from west to east, and the Dove and Manifold which cut across the southwest corner. These rivers flow in steep-sided dales, at times 200 m below the adjacent plateau surface. A complex network of dry valleys is less deeply incised into the plateau. The plateau is often quite sharply differentiated from the surrounding outcrop of Millstone Grit on the northwest and east, and from the Triassic rocks on the south. There are several convergent reasons for this, but they all spring from the coincidence of this area of limestone with a stable block (the Derbyshire Block-Kent 1967) across which shelf sedimentation occurred. The present area of limestone outcrop coincides closely with this stable shelf, and marginal dips around the dome are in part the original slope of the sediments of the shelf edge. Beyond them shales were deposited in the subsiding basins beyond the shelf, and this lithological contrast is reflected in the relief today, the shales forming a vale around much of the dome. Towards the close of the Lower Carboniferous the limestone massif was uplifted, and a period of erosion affected the crestal areas and the marginal zone with its reefs. Locally 100 m of limestone was eroded before the overlying shales were deposited. The same situation occurred locally in Triassic times, when the Triassic sediments were banked against the marginal slope of the limestones of the Derbyshire Block. This feature has been exhumed in places along the southwestern margin of the dome, as in the Weaver Hills, which rise steeply above the low Triassic plateaux to the south. Thus overall differential erosion has almost recreated the Visean shelf unit (fig. 12.1). Only in a few places, as at Great Hucklow and Eyam and south of Bakewell, do low dips at the limestone/shale junction suggest that the margin of the block still lies concealed by Namurian rocks.

The relationship of the covering Triassic to the Carboniferous rocks and structures across the southern end of the Pennines may give us some idea of the way in which the Pennines have developed. The sub-Triassic surface shows a good deal of relief largely unassociated with the Pennine structures, and while these structures pass southwards to the Midland horsts and are clearly seen at sites such as Breedon-on-the-Hill, the Pennines are terminated as a relief feature by a great east-west Triassic basin. We are left indeed with the impression that an internally complex upland block has been left upstanding (like the Midland horsts) by differential erosion of the surrounding Triassic rocks, but that as yet only the major streams have been able to make much impression on the more resistant elements of the Carboniferous outcrop. Further success here must at first seem fortuitous, for undoubtedly the greatest penetration is by the left-bank tributaries of the Trent, the Derbyshire Derwent and its tributary the Wye, the Dove, and the Manifold. Further north, where the Pennines are structurally both narrower and lower, the rivers have opened up the Coal Measures but made no more impact on the Millstone Grit outcrop than has been achieved further south. This suggests that the relief feature revealed as the surrounding Trias was cut away was a broadly rectangular horst, highest on the western side against the faults, and sloping gently eastward towards the westward-facing escarpment of the Magnesian Limestone. It cannot have had any close relationship to the pattern shown by the structural contours, unlike the closer relationship suggested by King (1962, 1876) in the Askrigg and Alston Blocks. We may suspect that this tilted horst was created in early Tertiary times and there may even have been some late Tertiary movement to tip up the western side, so ensuring a late start by the streams on the erosive exploitation of the Carboniferous structures (Johnson and Rice 1961).

With the exception of the col from Chapel-en-le-Frith to the Wye valley at Miller's Dale (the Dove Holes col (Jowett and Charlesworth 1929)) the Derbyshire Dome shows little effect of glaciation, although it almost certainly was ice-covered in the early glacial phases. Thus it shows the type of covered karst usual in the unglaciated areas of the British Isles, with a soil cover concealing most of the gently-sloping limestone surface (Pigott 1965). There are no limestone pavements, although erosion adjacent to steep dale slopes or the removal of the regolith by man can expose a deeply-weathered and highly irregular limestone surface. The soils are not entirely residual and include a proportion of Quaternary aeolian sediment (Pigott 1962). Steep valley (dale) sides show frequent outcrops of the more resistant limestone beds and in places quite high limestone cliffs occur. These reflect the ability of limestone to dispose of surface water quickly to the underground circulation, so reducing the effectiveness of surface mass-movement, allowing unusually steep slopes to form and survive. Caves and potholes are fairly common (Ford 1977), and the deep incision of the major streams on and around the limestone outcrop has given a low water table so that some deep potholes occur, notably Nettle Pot (*c* 120 m). The soil cover of the Derbyshire limestone gives relatively high carbon dioxide content to the percolating water and consequently fairly high carbonate values to the water issuing at springs after passage through the limestone (Pitty 1968b). However, the overall rates of solution are not unduly high as the rainfall is not particularly high for an upland area. The overall rate of denudation by solution has been discussed by Sweeting (1965) Dearden (1963) and Pitty (1968a).

The occurence of Tertiary sediments in the upper part of the pocket deposits within the Carboniferous Limestone (Walsh *et al* 1972) supports the view reached on geomorphological reasoning that the summit plane of at least part of the limestone outcrop is close to a Tertiary land surface. Most of the exhumation and associated incision and modification of the shelf sediments of the Derbyshire Block is clearly of late Tertiary and Quaternary age. The widespread plateau form and the rather close relationship between structure, lithology and relief around the margins of the dome have generally led to an interpretation that regards the limestone as relatively resistant to erosion except where rivers (some now dry) have been able to cut the deep dales. On this view, while not perhaps quite as resistant as the surrounding Millstone Grit outcrop, it is part of the surviving upland landscape of a Tertiary Britain. Pitty (1968a) has argued an alternative view, relating the present altitude of the limestone plateau to the time if has been exposed from beneath a

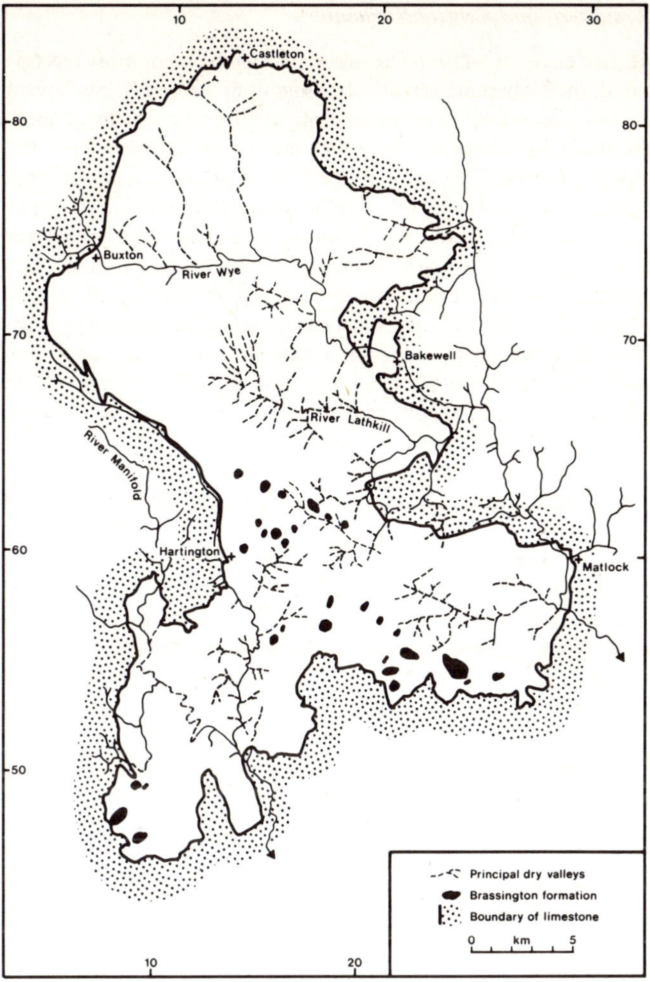

Fig 12.1 Dry valleys of the Carboniferous limestone and the Brassington Formation.
Redrawn from Ford 1978.

protective cover of shales. The higher parts have been most recently exposed, the highest areas still carry Namurian cappings. The lowest areas have been long exposed and are reduced to 100 m or more below the highest limestone levels by long-sustained denudation. The argument is difficult to test because it rests on an assumption that the limestone was once all at a level of about 500 m and at that level was fully covered with Namurian shales. The Tertiary pocket deposits suggest that part, at least, of the southern dome was stripped of its Namurian cover in the mid-Tertiary while the southern margin was clearly stripped of Namurian beds before the Triassic rocks were deposited. The argument is also rather circular in that incomplete Tertiary stripping of the Namurian cover would have left Namurian beds on some high, uneroded areas, just as they still survive Quaternary denudation in the Abney Low area east of Great Hucklow. It would not then be the duration of exhumation that determines the height of the limestone plateau at the present time but the varying success from place to place of Tertiary denudation. The view may seem old-fashioned, but there is evidence to support it. In addition, Namurian shales occur in the bottom of the Wardlow Mires basin, where the entire limestone sequence is present. These particular outliers survive in the lowest part of the landscape, suggesting appreciable relief on the sub-Namurian surface, a complication not considered by Pitty.

Subsidiary features of the limestone outcrop are the reef knolls (Wolfenden 1958). Towards the margin of the shelf major bioherms were built up and these marine limestones now prove far more resistant than the surrounding stratiform beds. They form hill features on the plateau and on the sides of some of the Dales (notably Dovedale), while others near Hassop are in the process of exhumation. Warwick has suggested that cave systems are preferentially developed in the reef limestones. He has also drawn attention to the dry valley system (Warwick 1963) and suggested that it developed on, and was superimposed from, overlying Namurian beds. As the main streams crossing the limestone and those around its margins were able to incise their valleys below the general plateau level, dewatering of the limestone led to progressive drying up of this valley system until only the very deeply incised rivers survived. In the case of the Manifold, which sinks for quite a long distance most summers, even some of these rivers are threatened by any further lowering of the water table. The present water table within the limestone has been lowered from its natural level over wide areas by artificial drainage. Long gently-inclined drainage tunnels known as 'soughs' (Rieuwerts 1966, 1969) were driven back from the major river valleys (especially the Derwent) to lower the water table and permit deeper mining without the expense of pumping. Many of these gravity-flow tunnels are still functioning, although most mining within the limestone has ceased.

The Millstone Grit and Coal Measures

In the high ground of the central Pennines from Kinderscout to the Aire gap, very gentle east-west folds help to guide the pattern of high grit plateaux cut by deeply-incised valleys that is so typical of the grit Pennines. All surfaces of low slope, whether on grit or shale, carry a considerable peat cover (Conway 1954). In many places this has been dissected by relatively recent erosion into a complex pattern of peat haggs with broad gulleys cut through to the solid rock and bordered by quite steep slopes in bare peat (Johnson 1957/8; Bower 1960; Radley 1962). Steeper slopes are frequently scarred by landslides where the massive grits capping the slopes have moved downhill with shear failures in the underlying shales (e.g. Johnson 1965c). In a number of areas these are still active or are readily reactivated by constructional activity, although generally the bulk of the movement occurred during the Pleistocene. Most of the valleys represent little more than the incision of the outward-flowing drainage, but Edale and the associated vale through to Castleton are broad valleys opening up the lowest shales of the Millstone Grit Series. Some of the northern valleys (e.g. Todmorden) have been greatly modified by glacial meltwater.

The second area of these younger Carboniferous rocks lies east of the limestone south Pennines. Here dips quite commonly range up to 15 or 20°, locally more where pitching folds complicate the general easterly dip. But overall dips are much lower, and the predominant impression of this area is of long cuesta features formed by grits and other coarse sandstones, characteristically with vertical rock outcrops wherever the drainage is deeply incised and the overall valleyside slopes are high. Such crags are best developed east of the Derwent which is deeply incised along the regional strike of the Millstone Grit. Stanage Edge on the Chatsworth Grits and Froggatt Edge on the Rough Rock are typical, and are bold enough and high enough to provide quite tough rock climbing. In detail these crags show blocks of immense size between the widely-spaced joints. Locally they culminate in tors and groups of upstanding boulders, argued by Linton (1964) to be the result of a two-cycle process. Certainly the well-rounded (woolsack) form of the tor blocks does not easily support Palmer and Radley's (1961) concept of a single-cycle frost-action origin. In any case, both exposures of deeply weathered regolith (albeit disturbed by solifluction) and Bunting's (1961) seepage hollows support the idea that felspathic grit (arkose) is readily modified by chemical weathering.

The broad cuesta and vale arrangement of the area east of the Pennines is modified by three factors. In the first place, the thickness, the wide separation of the bedding planes, and the coarseness of the grits and sandstones tend to decrease up the stratigraphical succession, so that the rocks become less and less resistant. There is an overall lowering of the surface, while the individual cuestas become rather less prominent, less likely to terminate in crags, and also usually less persistent along the strike. Secondly, the Millstone Grit outcrop in particular is complicated by a number of major structures, notably a series of quite steep periclines which swing from northwest to southeast across the scarp (fig. 12.2a and b). Some of these are related to pitching folds developed along the limestone margin (as at Matlock Bath), and all of them are likely to have been initiated in Hercynian times. Generally such folds have been inverted for at least part of their length, with inward-facing scarps and a deep anticlinal vale. As the famous example at Ashover shows, the full story of the evolution of such features, the adjustment of drainage to structure and the evolution of the structually-adjusted slopes is complex and difficult to disentangle (Sweeting 1946; Pitty 1967). Ashover, too, provides an excellent example of the areal limitation of such adjustment to structure for the flanking moors show surfaces of low slope bevelling the steeply-dipping beds which hardly affect the surface, despite the lithological contrasts present. At Crich inversion has only begun, and the core of Carboniferous Limestone forms a central domed hill, Crich Stand.

The third complicating factor in this area is the anomalous relationship of the main drainage lines to the regional structure. Alongside the close adjustment to structure that the Ashover anticline and the Don through Sheffield show, are numerous examples of insequent relationships. The Don meanders between Penistone and Sheffield (Linton 1956) show almost complete disregard for a strongly-banded geological surface although in detail they are related to intersecting faults: the Ashover stream below the inverted acticline vale turns westward to flow against the regional dip to join the Derwent at Ambergate. Most convincing of all, the Derwent flows southwards, broadly following the regional strike, yet cutting through a whole series of local folds as at Hathersage and Matlock Bath (Fearnsides 1932). The Derwent is a difficult element in the drainage pattern, for it is tempting to regard it as a subsequent stream, capturing the Pennine headwaters of an initial eastward drainage now represented by the Wye and the Noe. The height of the wind-gaps east of the Derwent requires that these hypothetical captures must have occurred at an early stage, and it may be that the Derwent has found itself trapped in the structures it crosses and has lost some of its initial adjustment to structure. The arguments have been reviewed by Linton (1956).

The predominantly, and probably original, eastward-flowing drainage of the southern Pennines has placed the main watershed along the western side of the Pennines. This means that the westward-flowing streams are short and steep, particularly as they quickly drop to the level of the adjacent Cheshire Plain. This western flank of the Pennines is also quite sharply folded with north-south fold axes and predominantly high dips. Some of the folds are inverted, usually only for rather short sections opened up by the headwaters of the Irish Sea drainage. The contrast with the east is greatest in the south where hog backs like the Roaches are found (Johnson and Rice 1961). Indeed the whole of the area from Glossop to Leek is characterized by relatively short north-south ridges separated by alternate north and south-flowing streams that link to break out westward to the Cheshire Plain. Towards the main watershed and further north dips fall off and the cuestas broaden, as at Axe Edge just south of Buxton. This western area is also complicated by the fact that it was invaded by the margin of the ice-sheet of the last glaciation (Jowett and Charlesworth 1929; Johnson 1965a), so that some of the cols have been lowered by glacial overriding, or more commonly by meltwater, increasing the dissection of the western Pennines. The high dips and associated faulting drop the Millstone Grit down below the Trias, so that there is normally no Coal Measure outcrop. The contrast is completed by the absence from the western margin of the Permian Magnesian Limestone.

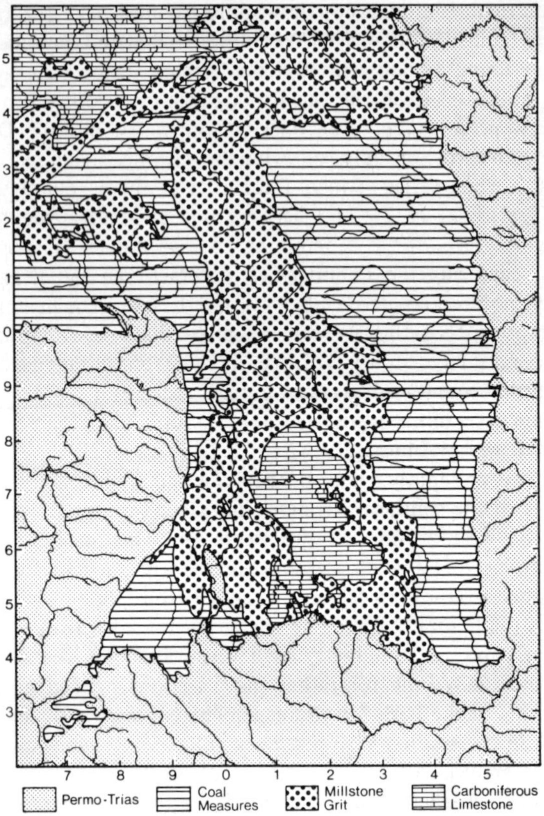

Fig. 12.2a Geology and drainage of the southern Pennines.

The Potteries

The Potteries coalfield and surrounding areas of north Staffordshire forms a distinctive region on the southwest Pennine margin. The Coal Measure syncline at the centre of the region is terminated abruptly on the west by the Red Rock fault which has a throw of about 1500 m and which brings the Triassic rocks of the Cheshire Plain against the Coal Measures. From this fault the drainage and the regional slope fall southeastwards towards the Penck valley, joining the Trent before it passes north of Cannock Chase. While the northern parts of the north Staffordshire coalfield rise to the level of the lower Pennine summits (Mow Cop, 335 m; Biddulph Moor, 335 m; The Cloud, 316 m) most of the area is much lower, sloping

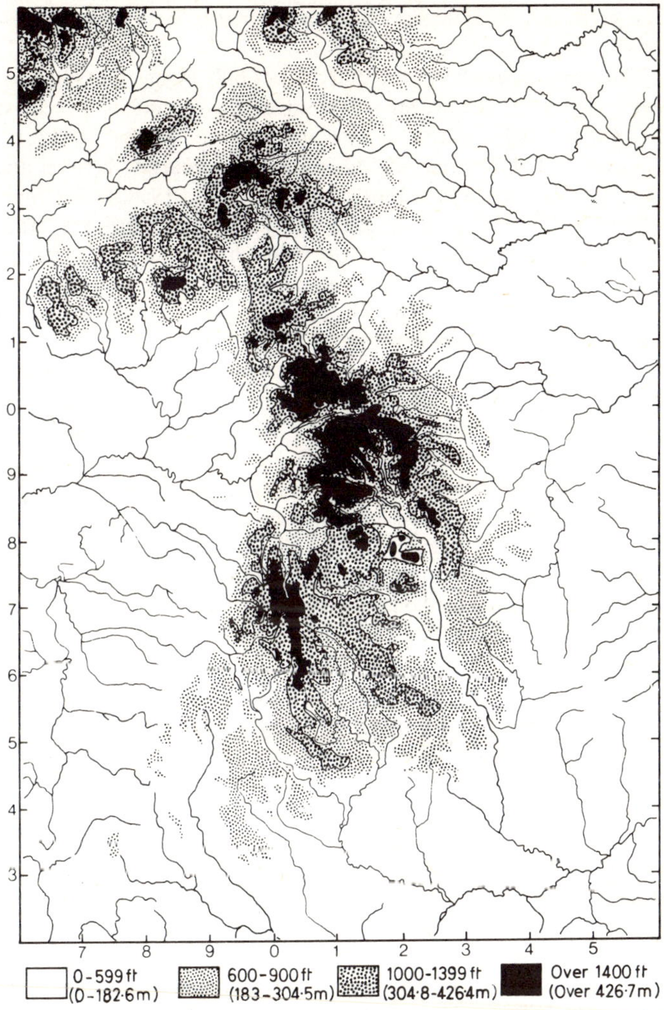

Fig. 12.2b Relief and drainage of the southern Pennines.

from 226 m at Alsager Bank north of where the M6 winds down to the Cheshire Plain, down to interfluves at about 137 m around Stafford. Further southeast the ground is higher on the pervious Bunter pebble beds, reaching 244 m in the highest part of Cannock Chase: indeed, the Bunter forms a northwest-facing scarp overlooking the Upper Coal Measures.

The northern part of the coalfield is not only higher, but has been deeply dissected. The process was completed by ice and its associated meltwaters, and a series of low coals diversified by meltwater channels form a complex and ill-defined water parting between the Cheshire Plain and the Trent drainage (Johnson 1965b). The most striking col is occupied by Rudyard Lake, but there are other features as, for example, south of Biddulph (Yates and Moseley 1958). At one stage a large part of the meltwater passed down the Churnet valley (King 1960) which was incised considerably as a result, while the low col between the Churnet and Trent headwaters (followed by the canal from Hanley to Leek) known as the Stockton Brook Gap indicates some rearrangement of the drainage pattern that must date from the retreat of the Irish Sea ice-sheet.

Yates (1956) has commented on the considerable degree of adjustment of the drainage to structure, with several of the streams occupying northwest-southeast outcrops of Etruria Marl. This is particularly true of the Fowlea Brook. He has suggested that the Trent has succeeded in integrating by capture a series of sub-parallel streams which originated on a sloping surface at about 215-225 m on its western edge above the Cheshire Plain. As he recognized remnants of this planation surface around Keele, he has named it the Keele surface. The surface is accordant across faults which displace Triassic beds, so it is a post-Triassic, not an exhumed sub-Triassic, surface. Whether a sloping subaerial surface would be associated with closely-spaced parallel streams seems unlikely, yet if they were already integrated by the Trent at that stage it seems unlikely that the parallel elements would have survived so well. Elsewhere such parallel streams have been interpreted as evidence of superimposition from a marine cover and this cannot be ruled out. There are some reasons for doubting marine (presumably Early Pleistocene) transgression so far west into Midland England, but it remains an hypothesis that cannot yet be rejected. Altitudinally the height of the Keele surface accords both with the well-established 'Calabrian' transgression of SE England and the Bierley Surface of Sissons (1954)

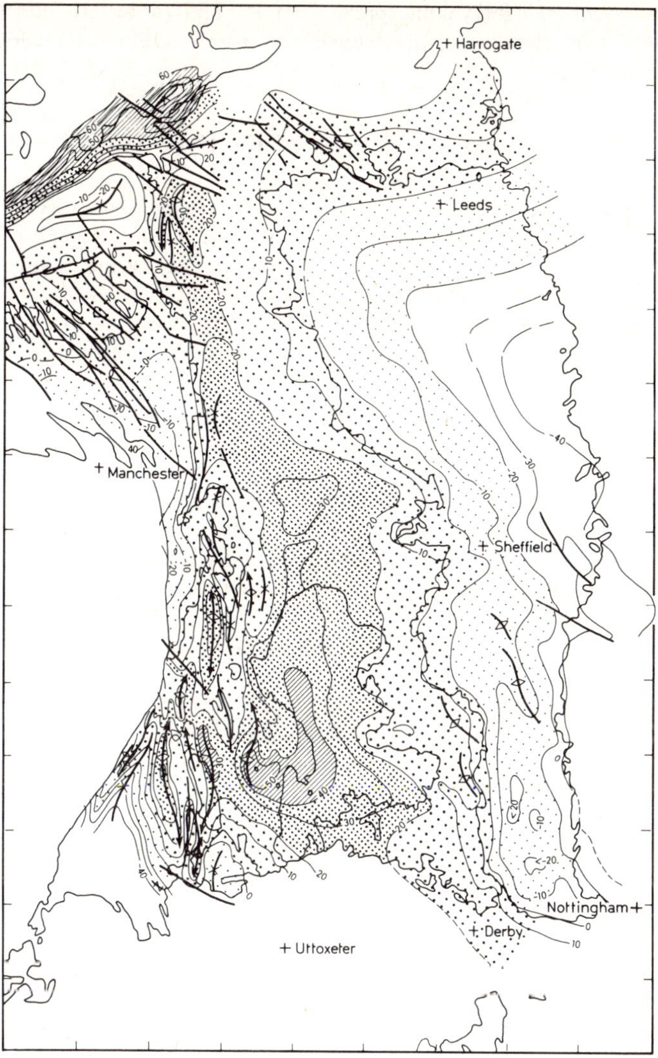

Fig. 12.3 Structural contours on the Rough Rock.

Contours on the Rough Rock drawn at 1000 ft (305 m) intervals.
Only the most important folds are marked.
This figure covers the same area as figures 12.2a and b.

also discussed in this volume (pp.87, 203). But until (or better, unless) correlated deposits are discovered, the matter will remain very speculative.

The Marginal scarplands west of the river Trent

The regular eastward dip of the Permian and Triassic rocks has at first sight relatively little expression in the relief of the area west of the Trent. Nearly everywhere along the eastern edge of the exposed coalfield the Magnesian Limestone caps a steep and regular scarp, but there are no comparable features to the east. The sequence of rocks gives rise to remarkable contrasts in land utilization, well shown in the published land-use maps of the area, but these changes are not accompanied by a regular sequence of scarp and vale. There are no doubt many secondary reasons which remain to be understood, but the principal cause seems to be the rarity of clay beds between the resistant rocks. The Magnesian Limestone is succeeded first by the Bunter Sandstone and then by the Pebble Beds, so that this highly porous, resistant formation fails to make a continuous scarp. In their turn the Pebble Beds are overlain by the lower parts of the Keuper Beds, the Waterstones, and in the absence of intervening clays, these again do not usually form a scarp feature. It is only the relatively weak skerry bands in the overlying Keuper Marl that make a relief feature readily traced on the 1:50000 map, and it is significant that it is on the clay vale to the west of this scarp that one of the few important subsequent streams of this area is to be found.

The Magnesian Limestone everywhere makes a feature except for a short distance west of Mansfield. Here the rocks from the Coal Measures on the west to the Bunter Pebble Beds on the east form a single relief unit which Swinnerton (1937) called the Mansfield Plateau. It has been suggested that the transgressive plateau surface, together with the superimposed drainage of the area, gave some evidence for former marine planation at about 185-200 m O.D.

Key: 1 Main (lower) Magnesian Limestone scarp
2 Upper Magnesian Limestone scarp
3 Bunter Pebble Beds scarp
4 Keuper Waterstones scarp
5 Bluffs at margin of the Trent trench (glacial margin route).

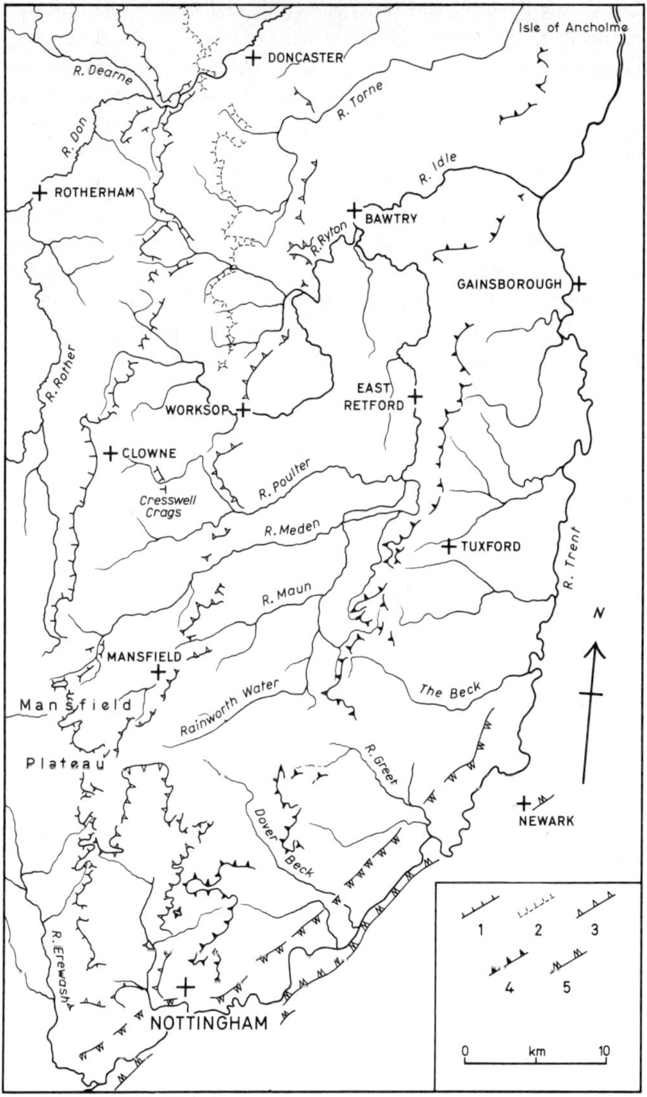

Fig. 12.4 Scarps east of the Pennines. (Key opposite.)

No part of the Permo-Trias outcrop of the northeast Midlands rises above this level, in fact lower surfaces have usually been developed at the expense of this local summit level, which is restricted to the area around Mansfield. Even in the Mansfield Plateau these lower surfaces are widespread; for example the erosion surface which completely bevels the Magnesian Limestone-Coal Measure junction near Teversall Colliery at 160 m.

The drainage of the Manfield Plateau must owe its preservation to the porous rocks across which the streams flow; in the absence of clay beds subsequent streams have failed to develop. There has been some integration of the parallel streams, for while these are only a mile or two apart in their upper courses they combine to form rivers some three miles apart. In some cases evidence of the capture remains, thus the Vicar Water, tributary to the river Maune, must once have flowed down the now-dry valley from Clipstone Forest to Rufford Abbey. It is likely that all these streams once continued eastwards across the Keuper skerry scarp, but they are all now tributary to the river Idle which flows, in the upper part of its course, on the Keuper Marl.

At the margin of the Mansfield Plateau there is frequently quite a sharp descent to lower levels, and the break below the 185-200 m surface appears to be one of the few important breaks in the sequence of planation surfaces found east of the Pennines. To the south, the crest of the Magnesian Limestone drops sharply west of Hucknall Torkard from 165 m to 130 m, and a similar break occurs on the Bunter Pebble Beds to the northeast of Newstead Abbey. To the north of Mansfield the plateau surface is preserved on the Magnesian Limestone as far north as Bolsover, the cuesta reaches 187 m between Glapwell and Palterton. There is then a steady fall to a lower level, and to the north of Clowne the maximum altitudes are 150-152 m, while to the northeast of Mansfield the Bunter Pebble Beds do not reach even 120 m.

The Magnesian Limestone escarpment follows a south-north course from Nottingham to Knottingley. Although the limestone itself thickens steadily to the north, the scarp-face is fairly constant in height, while its absolute height falls, reaching only 110 m between the Don and the Aire. In the south the cuesta is not always a simple feature, for strike faults cause inliers of the Red Marls which are accompanied by short vales and minor scarps. It is a feature such as this which forms Cresswell Crags. The Red Marl outcrop here is drained by two streams, one of which rises quite high on the main dipslope at Clowne. They unite to flow through a narrow water-gap in the limestone to the east. Caves on the gorge sides were once occupied by Upper Palaeolithic people, responsible for the important Creswellian culture (Armstrong 1931, 1939). The main scarp of the limestone in the south is seldom broken by watergaps. Only in the Mansfield Plateau, where the headwaters of the Meden and Maun rise on the Coal Measures, is the scarp broken, and here the gaps are at a high altitude, and the watershed on the Coal Measures to the west of the gaps is everywhere above 150 m.

Although to the north the water-gaps are more frequent, they have similar characteristics. With the exception of the Don and the Aire, none of the many streams which rise on the Coal Measures and flow eastwards across the Magnesian Limestone outcrop drains more than a few square kilometres of the Coal Measures. Thus the Chesterfield canal, utilizing the Ryton water-gap, the most southerly of this northern series of gaps, crosses the Magnesian Limestone at about 110 m, but has to tunnel for nearly 3 km through the waterparting on the Coal Measures to the west. The river Don shows signs of superimposition across the Magnesian Limestone, for the river does not utilize the shortest route across the outcrop. This occurs a kilometre or two south of the Don gap where, as the result of faulting, only a mile of Upper Magnesian Limestone separates the Coal Measures from the Upper Permian Marl. This upper limestone, which occurs north of Worksop, is separated from the lower by the Middle Permian Marls and so builds a second feature to the east of the main escarpment. This second scarp is very broken between Tickhill and the river Don, but between the Don and the Aire, the secondary scarp is more continuous and the vale of the Middle Marls is followed by the former South Yorkshire Junction railway from Sprotborough to Upton (Linton 1956).

Although there is no continuous feature marking the oncoming of the Bunter Pebble Beds, the Upper Permian Marls are locally followed by a low vale, sometimes accompanied by a short stretch of subsequent drainage. The most important of these subsequent streams is the river Leen, which flows southward for 15 km from Newstead Abbey to the river Trent, just above Nottingham. Its strike valley is highly asymmetrical, with a wide dipslope on the Magnesian Limestone to the west, and a steep scarp capped by Bunter Pebble Beds to the east. The headward development of the Leen has resulted from the springs occurring at the Bunter-Upper Permian Marl junction, but the stream has no doubt also been aided by the considerable thickness of the Bunter Sandstone in this area, a bed which is usually considerably less resistant than the overlying Pebble Beds. It is likely that the Leen cut back from the Trent rapidly, for there are no low wind-gaps on the east of the stream to mark the course of the former drainage. On the other hand, the streams dissecting the Mansfield Plateau have lowered the water table so far that the head of the Leen has clearly not been receding actively for some time, and there is probably little chance of it ever capturing the Rainworth Water. In any case, the upper part of the Rainworth Water valley, separated from the Leen by the Mosley Hills, is already dry. Linton (Wooldridge and Linton 1955) described similar cases of the stagnation of formerly active streams in the porous Wessex chalk.

On the narrow interfluves that separate the eastnortheasterly-flowing streams of the Manfield Plateau, the Bunter Pebble Beds form the lower, eastern portions of each ridge. In some cases the ridge rises a little at the Bunter outcrop, elsewhere there is no feature. Locally, as at Welbeck Park, a short stretch of subsequent drainage on the Upper Permian Marls gives a well-defined scarp on the Bunter rocks. On the other hand, there can be no mistaking the change in topography as the Bunter rocks are reached. Although both the Magnesian Limestone and the Bunter rocks are porous, they give rise to contrasted topography. The Magnesian Limestone is characterized by gently-sloping plateaux broken by narrow dry valleys that frequently show right-angled bends. The valley sides are cut sharply below the plateau, reflecting the bedding of the limestone. There are no such sharp breaks on the Bunter rocks, for they are practically devoid of bedding. Changes of slope occur relatively gradually, resulting in a surface that is everywhere rolling, a succession of smoothly-flowing concave and convex slopes. This is emphasized by the complex pattern of minor branching valleys, so that every valley-side in its turn is broken into a continuous wave of minor slopes.

The high ground of the Bunter Pebble Beds east of the river Leen is continued around the Day Brook and into the northern part of Nottingham by the Keuper Waterstones. These underlie a plateau that is everywhere above 105 m and rises to just over 150 m at Ramsdale Park. To the northeast, the Waterstones feature is less important and is replaced by the scarp of the skerry bands which occur towards the base of the Keuper Marl. As both the Permian and Triassic rocks thicken northwards, the skerry feature is not parallel to the Magnesian Limestone scarp, but lies about 20° east of north, from the northeastern outskirts of Nottingham to a kilometre or two west of Tuxford. The main skerry 'belt' forms a resistant cap-rock over the lower part of the Keuper Marl, and some of the steepest slopes in the area are found on the Keuper Marl slopes below these skerries.

Although the Keuper (skerry) scarp is quite a high and important feature, it is broken by a series of gaps. In the south these occur where tributaries flowing southeastwards to the river Trent drain part of the area west of the scarp; in no case is the area large, nor has a subsequent stream of any lenth developed on the lowest part of the Keuper Marl. To the north the scarp is almost unbroken, but the Idle (above Retford) and its continuation upstream to the Rainworth Water drain what, in part at least, is a subsequent vale. The Keuper scarp to the east is broken by a series of wind-gaps, marking the former continuation of the eastnortheast streams of the Mansfield Plateau. These particular streams are unlikely to have flowed across this cuesta for some considerable time, but it is likely that some of the deeper gaps carried short streams until more recently. Thus the gap followed by the Chesterfield and Lincoln railway west of Tuxford probably carried the short stream that rises at Wellow for some time after the river Maun was captured by a tributary of the river Meden.

The southern part of the cuesta of the Keuper Marl shows considerable signs of rejuvenation. The southeasterly flowing streams formerly joined the river Trent further to the southeast at the time that it flowed through the Ancaster Gap (Clayton 1955, Straw, 1963); with the Wolstonian diversion of the Trent to its present course across the Keuper Marl dipslope, these streams found themselves with greatly shortened courses to the river Trent. In addition, the Trent itself had cut down to a lower base-level. The result is that knick-points have worked back up these streams, and they have attacked ever more vigorously the readily-eroded Keuper Marl. The main streams are now adjusted to the changed conditions, but their tributaries, and also the shorter and smaller streams that flow directly to the Trent are still in a rejuvenated condition. These steep and vigorous streams are known locally as 'dumbles'. Around the head of some of them the dissection of the skerry belt and the marl slopes that the skerries protect is particularly complex.

The Midland horsts

East Warwickshire horst

By no means all the Midland inliers of Palaeozoic rocks really deserve the term horst, although it is a convenient way to refer to them. However, the east Warwickshire coalfield is fault-bounded and generally does rise above the surrounding Triassic rocks. It can be taken as typical of these features and will be described at greater length than the other inliers (Eastwood *et al* 1923). Its maximum dimensions are 13 km from east to west, by some 40 km north to south. The bounding faults, which have throws in the order of 200-300 m, bring the Keuper Marl against Upper Coal Measures and in places against Pre-Cambrian igneous rocks. Relief differences between the surrounding Triassic rocks and the Palaeozoic inliers are not large, but typically the Trias lies below 100 m while the Palaeozoic rocks commonly rise above 100 m. The highest point of the east Warwickshire coalfield is at 185 m north of Meriden, and the extensive outcrop of the Keele Series (Upper Coal Measures) reaches 177 m some 12 km further north near Bentley.

Structurally the east Warwickshire coalfield is a syncline, its axis running north-south through the eastern part of the Palaeozoic outcrop. The eastern flank of the syncline has dips of 10-12°, and in quick succession belted outcrops of Lower Coal Measures, Cambrian (Stockingford) Shales, Cambrian (Hartshill) Quartzite and Pre-Cambrian igneous rocks (Caldecote Series) occur. While there is local expression of these beds in the relief, it is on a subsidiary scale, and at the regional level the internal structure of the horst finds no expression in the relief. In the centre of this eastern syncline there is something like 500 m of Coal Measures. The syncline is broken westwards by Middle Coal Measures brought towards the surface — the dip on this limb reaches about 6°. The western part of the horst consists of almost horizontal Keele Beds which dip very gently eastwards along the western edge of the inlier.

As has already been noted, the relationship of the east Warwickshire horst to the regional drainage suggested adjustment of that drainage to structure over a long period of time. There may have been some Tertiary movement along the faults, but if so it has been overshadowed by surface lowering brought about by the direct ancestors of the present drainage pattern. In detail this conclusion is supported by the relationship of the surrounding rivers Blythe, Thame and Anker to the inlier. The Blythe, along with the Hatchford Brook, the river Cole and the Thame above the Blythe junction have all established drainage basins on the Triassic rocks. The eastern tributaries of the Blythe have opened up a number of small basins that drain the western flank of the inlier, but only the Burne has succeeded in integrating a basin of any size within the Palaeozoic outcrop. Success is probably related to the establishment of a basin on overlying Triassic rocks, and maintenance of the main water-partings after the streams unroofed the Palaeozoic horst. In the north the swinging meanders of the Anker are unchanged as they come from the Trias, across the northernmost spur of the horst and back on to the Trias again, and they, too, suggest relatively recent superimposition from a continuous cover of Trias and argue against any appreciable late-Tertiary or Quaternary movement along the bounding faults.

South Staffordshire-Lickey Hills horst

The rather simple structure and outline of the east Warwickshire horst may be contrasted with the next inlier to the west (Whitehead and Pocock 1947; Whitehead *et al* 1927). This includes the Black Country region, based on the coal seams of the south Staffordshire coalfield, and the high southern edge on the coarse sandstones and breccias of the Upper Coal Measures (Enville Beds). The Palaeozoic rocks here are more sharply faulted and folded than in the east Warwickshire horst, and this is best seen in the steep dips of the inliers of Silurian (Wenlock) limestone at Dudley Castle and Wrens Nest. The bounding faults are best seen along the western margin of the horst, from Wolverhampton to the Clent Hills; on the east the very thick accumulation of Upper Coal Measures means that despite throws of nearly 300 m, the faults bring Enville Beds against Keele Beds, and the Triassic margin is rarely marked by a fault. Here, too, is evidence that these inliers have been revealed gradually by the stripping away of a Triassic cover.

A combination of more resistant rocks and a position on the central water-parting of the British Isles means that the altitudes on this inlier are appreciably higher than in east Warwickshire. The ground reaches 267 m west of Smethwick on intrusive dolerites within the Etruria Marl, while southeast of Birmingham the Lickey

Hills reach 291m and the Clent Hills 316 m. Here steep slopes lead down towards the Severn drainage and at Clent the Keuper Sandstone across the fault lies between 125 and 160 m. The significance of lithology rather than age or structural relations is shown in the northern edge of this horst in the Cannock area. The northern prolongation between Cannock and the fault west of Brereton is overstepped by the Bunter Sandstone outcrop of Cannock Chase. Much of the Coal Measure outcrop here is thickly covered by till, but a summit surface at about 235 m forms the highest ground on the Triassic and Palaeozoic rocks. Along the actual junction the Bunter forms an appreciable escarpment, as in Beaudesert Park where the Coal Measures at 200 m is overlooked by the Bunter which rises to 237 m in the Rawnsley Hills. It is characteristic of the differential erosion that this represents, that the Bentley Brook heads on the Coal Measures and turns west through a water gap into the Bunter.

Wellington and Enville horsts

The river Severn from the lower limit of the Ironbridge gorge to several kilometres below Bridgnorth flows along the Coal Measures-Triassic junction, separating rather neatly the Welsh borderland from the Midlands. Above and below this reach it cuts across the Carboniferous rocks separating off three northeastward projections which form horst-like spurs from the Carboniferous and Old Red sandstone outcrop of the Welsh borderland. The northernmost is almost entirely buried beneath drift, although, just northeast of Shrewsbury, Longmyndian rocks form Haughmond Hill (152 m). The fault bounding this on the northwest lines up with the Red Rock fault along the northwest side of the north Staffordshire coalfield. The next horst to the southeast includes the isolated hill of the Wrekin, where resistant Pre-Cambrian (Uriconian) rocks (Pocock *et al* 1938; Earp and Hains 1971) form a steep-sided ridge which rises to 406 m. Silurian rocks also outcrop north of the river Severn within this unit, but as with most of the Midland horsts, the Coal Measures form the major outcrop of Palaeozoic rocks. The steep-sided gorge at and below Ironbridge is cut in Coal Measure sandstones and shales. The predominant dip is southeasterly, or easterly, so that the productive (Middle) Coal Measures are succeeded to the east by the Etrurian Marl, Keele Beds and Enville Beds of the Upper Coal Measures.

The last of this group of horsts is the wedge-shaped Enville horst

(Whitehead and Pocock 1947). It is bounded on the west by a fault trending about 35° east of north, and on the east by a fault running no more than 15° east of north. Consequently the Palaeozoic outcrop narrows from 8 km near the Severn valley to only 4 km where the Palaeozoic beds are finally overstepped by the Trias. Apart from a small area of Old Red standstone down towards Kidderminster in the south, almost the entire outcrop northeast of the Severn consists of Keele or Enville Beds. Southwest of Enville itself the ground rises to 204 m.

South Derbyshire and Charnwood

The remaining areas of Pre-Triassic outcrop are rather different from the stripped horsts so far described. In the south Derbyshire coalfield even the highest part of the Triassic/Palaeozoic contact is close to the present land surface, so that the bounding faults are almost everywhere concealed by Triassic rocks (Sylvester-Bradley and Ford 1968). Parts of the Coal Measure outcrop have been revealed by erosion, but only the northwest-southeast fault from Tickhall to Thringston shows a marginal relationship between the Palaeozoic and Triassic rocks at all like that seen elsewhere in the Midlands. Further inspection, however, shows that even this boundary is complicated by a broad outcrop of Millstone Grit and some underlying Carboniferous Limestone that extends 5 km to the Trent between Melbourne and Stanton. In addition, extensive remnants of the cover of Keuper Sandstone survive above the Coal Measures both north and east of Ashby-de-la-Zouch.

Northeast and east of the south Derbyshire coalfield, irregular hills have been exhumed from beneath a Triassic cover. In places, as at Breedon-on-the-Hill or Breedon Cloud, the hills are of Carboniferous Limestone, but most characteristically they are the Pre-Cambrian volcanics of Charnwood Forest. Here a scatter of hills of different sizes, some of them capped by rocky tors, rises through the sheet of Keuper Marl. The highest, Bardon Hill, reaches 278 m, and many reach 200 m. While the greatest concentration lies in Charnwood itself, to the northwest of Leicester, outliers occur far afield, as at Enderby and Croft, 16 km south of the centre of Charnwood itself.

The summits of several of the Charnwood hills are capped by tors, no doubt first prepared by deep weathering of unknown (possibly

Tertiary) age, but finally stripped by frost action in the Quaternary (Ford 1967, 1968). As a result we have a remarkably polygenetic relief in Charnwood, with exhumed sub-Triassic wadi-like valleys, alongside block-strewn slopes of periglacial origin. On the resistant rocks steep slopes suggest a relief energy that is hardly matched by the modest altitudes that are generally around 200 m.

The Jurassic scarplands

The southeast Midlands can be subdivided on a topographical basis into four zones:

(1) To the west, the Inferior Oolite and the Marlstone outcrops form the watershed between the Trent or Severn drainage and those rivers which flow more or less directly to the Wash. The more resistant rocks have been greatly dissected and the area is characterized by a highly irregular scatter of outliers which range in size from the conical hills of Whatborough or Robin-a-Tiptoe west of Oakham, to the deeply scored plateau of Cold Ashby, east of Rugby. The scattered glacial deposits are generally confined to the hills, except between Leicester and Rugby where they are very thick and infill a preglacial Soar valley.

(2) To the east and southeast of this dissected zone lies an area of broad plateaux separated by the main river valleys. Although the plateaux are based on the Great Oolite in the west, they are principally built of glacial till and vary only in the stage to which they have been dissected. The plateaux were once densely forested, and although now largely cleared, retain regional names such as Salcey Forest and Yardley Chase.

(3) Still further southeast beyond this zone of drift plateaux lies a more complex zone where the glacial deposits are more deeply dissected and Upper Jurassic and Lower Cretaceous rocks from isolated hills and cuestas. Most important of these is the Lower Greensand cuesta north of Leighton Buzzard. This area includes the limit of the advance of the Wolstonian ice-sheet and as a result the drift deposits are more closely controlled by the topography than they are to the north.

(4) The northern edge of the Jurassic plateaux lies around Melton Mowbray and south of the Vale of Belvoir. Here a series of well-developed scarps is closely related to Tertiary folding (fig. 12.8). In

particular the Marlstone scarp north of Melton Mowbray forms an impressive feature south of the Vale of Belvoir, excavated by glacial erosion of the Lias clay outcrop (see Straw, in this volume, p.58).

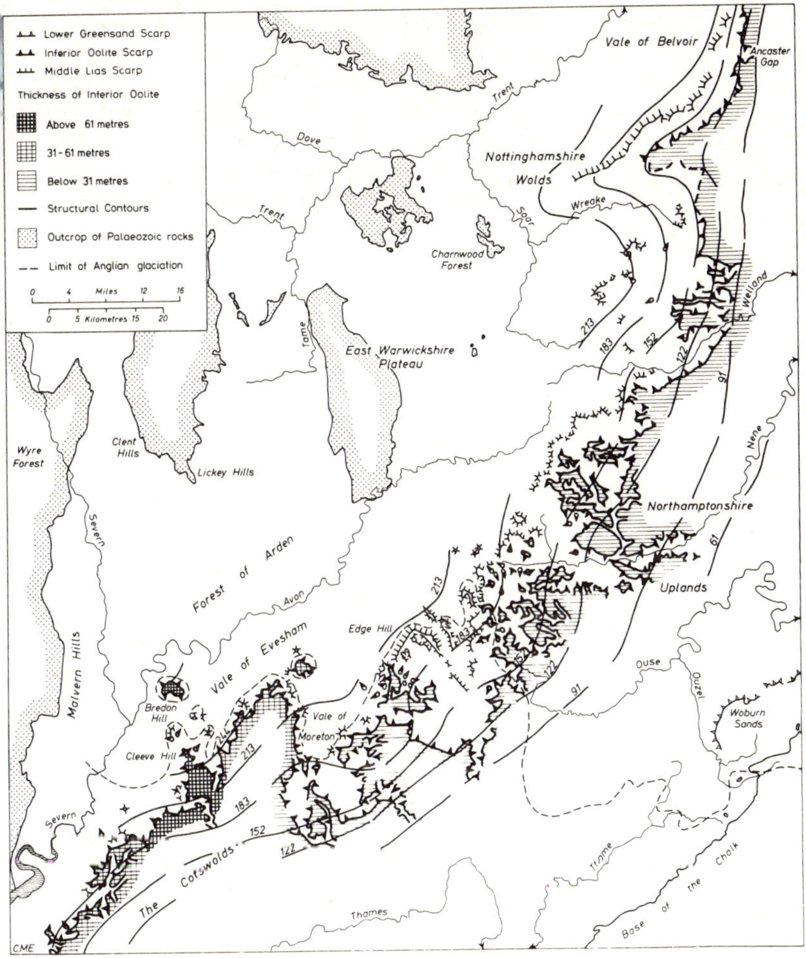

Fig. 12.5 The Jurassic scarplands and plateaus.

The western zone

The higher, dissected, western zone forms the main hill belt of the Jurassic scarplands. Altitudes are highest in the north where both the Marlstone and the Interior Oolite are capped by fairly thin remnants of boulder clay. Here the summits lie below 210 m. It is this central zone that includes the main routes across the Jurassic scarplands. These routes do not cross in gaps, but follow the upper tributaries of the Nene and Welland and then cross the narrow ridges dividing these valleys from the Avon or Soar valley to the west.

The plateaux of this western zone are partly related to the low dip of the rocks. In addition, there is evidence that in places the plateaux are erosional features, bevelling the rock outcrops (Kellaway and Taylor 1953). Although such planation surfaces may be recognized locally, they cannot be fully described until an accurate drift map has been made for his area. Thus to the northwest of Northampton there is some evidence for a surface at about 180 m, no doubt related to the similar feature west of Oakham. The 600 ft (c. 180 m) closed contours are drift-free, as is the ridge lying at 180-6 m east of Naseby. It might appear that the surface is also found on the Marlstone east of Sibbertoft, but most of this ridge is thickly covered by chalky boulder-clay.

The lower surface west of Northampton cuts across the Northampton Sand, and at Little Brington the Inferior Oolite scarp is cut by this surface at about 140 m. The Marlstone outcrop flanking the Nene west of Northampton is a largely drift-free plain at a little above 90 m. The surface continues on to younger rocks to the east and south, and this suggests that it is an erosional, rather than a structural feature. From the surface rise remarkable small, conical hills, outliers of the Northampton Sand. The highest of these, Gawburrow Hill, rises to over 137 m. The morphology of the area

---

Fig. 12.6 The geomorphology of the area between Rugby and Towcester.

This figure shows a characteristic section of the highly dissected Jurassic plateaux.

Key: 1 Marlstone (Middle Lias) scarp
2 Inferior Oolite scarp
3 Great Oolite scarp
4 Till
5 Glacial gravels (with some associated till)
6 Stream courses initiated on till surface.

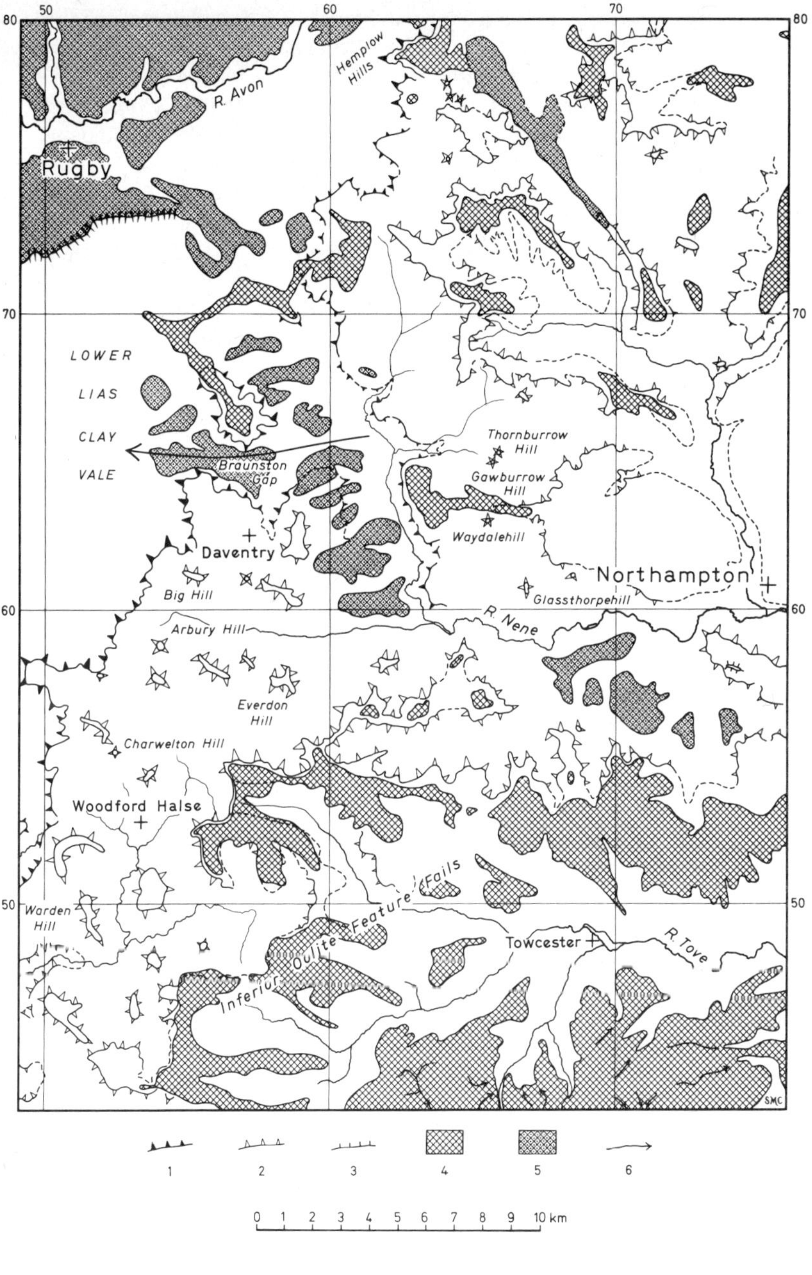
50
60
70
80
R. Avon
Hemplow Hills
Rugby
LOWER LIAS CLAY VALE
Braunston Gap
Thornburrow Hill
Gawburrow Hill
Waydalehill
Daventry
Northampton
Big Hill
Glassthorpehill
Arbury Hill
R. Nene
Everdon Hill
Charwelton Hill
Woodford Halse
Inferior Oolite Feature Fails
Warden Hill
Towcester
R. Tove
SMC
1
2
3
4
5
6
0 1 2 3 4 5 6 7 8 9 10 km

west of Northampton is shown in Figure 12.6. It will be noticed that midway between the Nene and the Welland, to the east of Rugby, the Northampton Sand outliers are hardly a kilometre from the Marlstone scarp. This distance increases rapidly both to the north and south, a pattern that is repeated both between the Welland and the Wreak, and between the Nene and the Ouse.

The important northern tributary of the river Nene, followed by the railway line and M1, generally known as the Watford stream, seems to have suffered diversion during the Wolstonian glaciation. The valley, unlike others in the area, contains a series of glacial sands and clays, and these also fill a deep col at Braunston Gap, where the Grand Union canal crosses the present watershed in Braunston tunnel. The east-bank tributaries of the Watford stream flow towards the Braunston Gap, and it is possible that the Watford stream has only flowed to the Nene following the blocking of the gap by glacial deposits.

To the south of the Nene, the outliers of Northampton Sand show a change in shape. Characteristically they are narrow ridges, considerably longer than they are broad, yet they rarely reach much more than 1.5 km in length. They retain a narrow plateau surface, and while those in the west rise above 215 m, the remainder tend to have summits at about 180 m. The cap-rock on these ridges protects constant, and fairly steep, slopes on the underlying clays. The ridges are more continuous to the east, but even here, they are locally very narrow, and it is clear that in time they will break up into a line of smaller hills.

The landscape is even more dissected within the Cherwell basin, and the divide between the Cherwell and the Tove has locally shifted eastwards as a result of river capture by the Cherwell. The chalky boulder-clay does not normally enter the Cherwell basin, and the lobe of it southeast of Woodford Halse that is now drained to the Cherwell appears to locate a recent capture of part of the Tove drainage. The capturing stream cut back through the Northampton Sand scarp near Eydon. To the east of the upper Cherwell, within the Ouse-Tove basin, the topography is related to the boulder-clay cover, and falls in the second zone of the southeast Midlands.

The boulder-clay plateaux

East of the uplands associated with the Marlstone and Inferior

Oolite outcrops, the landscape consists of a series of broad plateaux, dissected along their margins and separated by the trench-like valleys of the major rivers. Although some of the plateaux are formed of solid rock, many are built almost entirely of boulder-clay. Only locally are cuesta forms developed, although north of the river Nene the Great and Inferior Oolite form wide dipslopes that are only lightly covered in glacial drift.

The evolution of the drainage pattern of the southeast Midlands is little understood, largely because of the thick cover of drift. A reconstruction for the Corby-Stamford area, mainly based on the borehole evidence available for the ironstone area (Kellaway and Taylor 1953), suggests that the drainage there has evolved from a series of closely-spaced west-east consequent streams. These streams have been disrupted by the preglacial development of subsequents, cutting back along the clay outcrops. Part of the area is shown in Figure 12.6. Many of the wind-gaps are now drift-filled, and only locally, as north of Stamford, have they been 'exhumed' by the removal of boulder-clay.

The area north of Stamford is dominated by the valleys of three supposed subsequent streams, the North Brook, the Glen and the Braceborough rivers. They flow in remarkably straight and parallel courses, deeply trenching the easterly-sloping plateau. The higher, northern portions of the interfluves between these streams are capped with finger-like remnants of boulder-clay. This boulder-clay does not descend far below 30 m anywhere along these valleys, and west of Bourne it caps a marked bluff, at about 45 m. To the east, along the Fen margin, lie two other deposits: in the south a broad spread of gravel, and in the north a low, cuesta-like ridge of boulder-clay. It has been suggested (Clayton 1957) that this boulder-clay belongs to an early-Devensian advance. The gravel to the south is mapped by the Geological Survey as Fen Gravel, and some of it is marine. However, the main deposits of the gravel are broad, delta-like spreads associated with each of the rivers entering the Fens.

Between Stamford and Peterborough, the Welland-Nene divide is a broad, largely drift-free plateau with a gentle slope to the east that is largely a true dipslope. The col at Southorpe, aligned with the Oundle reach of the Nene, probably functioned as a meltwater channel at a time when an ice-sheet occupied the Fens. To the south of Peterborough, stretching for some 50 km to the Ouse, is one of the most extensive boulder-clay plateaux of the East Midlands. The underlying rock is principally the Oxford Clay, but the present landscape is entirely postglacial in origin. The boulder-clay surface rises to 103 m at Sharnbrook cutting on the Midland railway, but for many kilometres southeast of Oundle reaches only 65-75 m. The major streams of this plateau, such as the Alconbury Brook and the river Kym, flow almost exactly southeast, and they receive a series of closely-spaced tributaries from the southwest. The most likely hypothesis would seem to be that the initial consequent drainage towards the northeast has been disrupted by the growth of the southeast flowing subsequents. This is supported by the general slope of the surface of the boulder-clay which is to the east or eastnortheast. The subsequent streams appear to owe their success to the impermeability of the underlying Oxford Clay, for there is an important springline at the base of the boulder-clay. The dissection of this plateau has proceeded furthest in the south where the height of the boulder-clay surface (and, to a certain extent, of the drift base) is greatest.

Southwest of Sharnbrook cutting, the boulder-clay broadens again through Yardley Chase and Salcey Forest. It then narrows where the river Tove flows southeast towards Wolverton, and at Roade cutting, where the railway line to Crewe crosses into the Nene basin, the drift is now little more than a kilometre wide. The short streams dissecting the drift plateau edge have eroded it into a characteristic lobate form, and the stage reached in the dissection of the plateau is remarkably uniform around the whole of the plateau margin.

Further detailed description of the drift plateaux to the west and south would merely be tedious. They are very similar to the examples already described and vary only in the pattern of the drainage and the stage to which dissection has proceeded. The Tove-Ouse divide carries Whittlewood Forest (maximum elevation 156 m), south of the Ouse and west of the Ouzel is Whaddon Chase (153 m). The Whittlewood Forest plateau has a radial drainage pattern, while Whaddon Chase has many parallel northeasterly flowing streams with very little development of subsequent drainage. This survival of the consquent streams is probably related to the fact that few streams have cut down to the base of the boulder-clay. An exception is the subsequent Loughton stream, northnorthwest of Bletchley.

The final area in this zone of drift plateaux is the more deeply dissected country around Kettering, between the Nene and the Welland. The boulder-clay here is thick, but the main streams have all cut valleys into the underlying rocks, and the drift survives as finger-like remanants on the intervening spurs. The ridge east of Corby continues to the east and broadens rapidly into a small plateau that is the nucleus of Rockingham Forest. The reconstruction of the subglacial surface in this area has revealed numerous drift-filled wind-gaps, evidence of a former west-east system of consequent streams. This stream pattern was disrupted prior to the glaciation of the area by subsequent development on the softer outcrops (Kellaway and Taylor 1953). The most recent important capture for which there is clear evidence is the capture of the upper Welland by the Gwash. The former eastward route of the Welland is marked by a drift-filled wind-gap with its floor just above 76 m south of Harringworth. Although former wind-gaps have been destroyed by later erosion, it is possible that the Oundle reach of the Nene is no older. There is no surface evidence of the drift-filled wind-gaps of the Corby area, and it is highly probable that such features occur elsewhere beneath the drifts of the East Midlands.

The southeastern margin

The marginal areas of the East Midlands inevitably show transitional features, some of which are typical of the regions beyond. To the northeast of Leighton Buzzard the Lower Greensand forms a cuesta, while to the south and west Upper Jurassic and Lower Cretaceous rocks cap a series of isolated hills. The maximum limit of advance of the ice-sheet occurs in this region and consequently the glacial deposits differ in character and form from those already described.

The Lower Greensand cuesta is not a continuous feature, although to the south of Bedford the line of the scarp is continued by a feature developed in boulder-clay. The limit of the Lower Greensand outcrop can be traced in the plateau behind this feature around Old Warden. It reappears east of the river Ivel at Sandy where it builds an impressive bluff. The scarp is then continued northeastwards to Waresley, where it again disappears beneath the boulder-clay. In a southwesterly direction the cuesta fails along the river Ouzel, partly because the Lower Greensand becomes thinner, and partly because it is buried beneath the Gault overstep. The Lower Greensand forms a wide and high feature immediately east of the Ouzel, but the streams on the backslope have exposed the underlying Jurassic clays, and the dissection increases to the east. In two places, at Husborne Crawley and at Ridgmont, the scarp is broken by obsequent streams. Parts of the backslope are covered in chalky boulder-clay which gives a landscape contrasting strongly with that of the Lower Greensand. The Shefford area, west of the upper Ivel, shows in miniature the remarkable contrasts between the Jurassic clay of the valley floor, the Lower Greensand valley side with dry valleys and heath vegetation, and the smooth, arable plateau of the boulder-clay.

Both the Ouse and the Ivel show wide river terraces at heights up to 18 m above alluvium. The wide break in the Lower Greensand escarpment west of Biggleswade is now obscured by boulder-clay; it is strongly contrasted with the well-developed cuesta to the east, bevelled by a smooth surface at about 67 m. The development of both the Ivel and the Ouzel has not so far been studied, and the two rivers pose many problems. The boulder-clay plateau east of Biggleswade covers a wide variety of rocks and the preglacial topography was very different from that of today. The drift surface is very level and consequently the drainage pattern is rather irregular.

A few kilometres west of Ashwell the river Ivel flows north through the Lower Greensand ridge to join the Ouse. Edmonds and Dinham (1965) identify a subdrift depression extending from near Biggleswade, northeast to Hatley, as the preglacial valley of a former trunk stream that took both Ivel and Rhee waters northeast towards Cambridge, but it could simply be a subglacial erosional feature. The present Ivel gap at Sandy is probably a glacial feature opened up by a combination of glacial erosion and the action of meltwater (Edwards and Dinham 1965) such that when the ice disappeared, the rivers found a northward, rather than a northeastward route, much in the manner of the Trent at Gainsborough.

Key: 1 Portland Limestone scarp
2 Lower Greensand (Woburn Sands) scarp
3 Lower Chalk (Totternhoe Stone)
4 Middle Chalk (Melbourne Rock) scarp
5 Upper Chalk (chalk rock) scarp
6 Till
7 Inferred glacial limit
8 Summit heights on till surface
9 Streams originating on till surface.

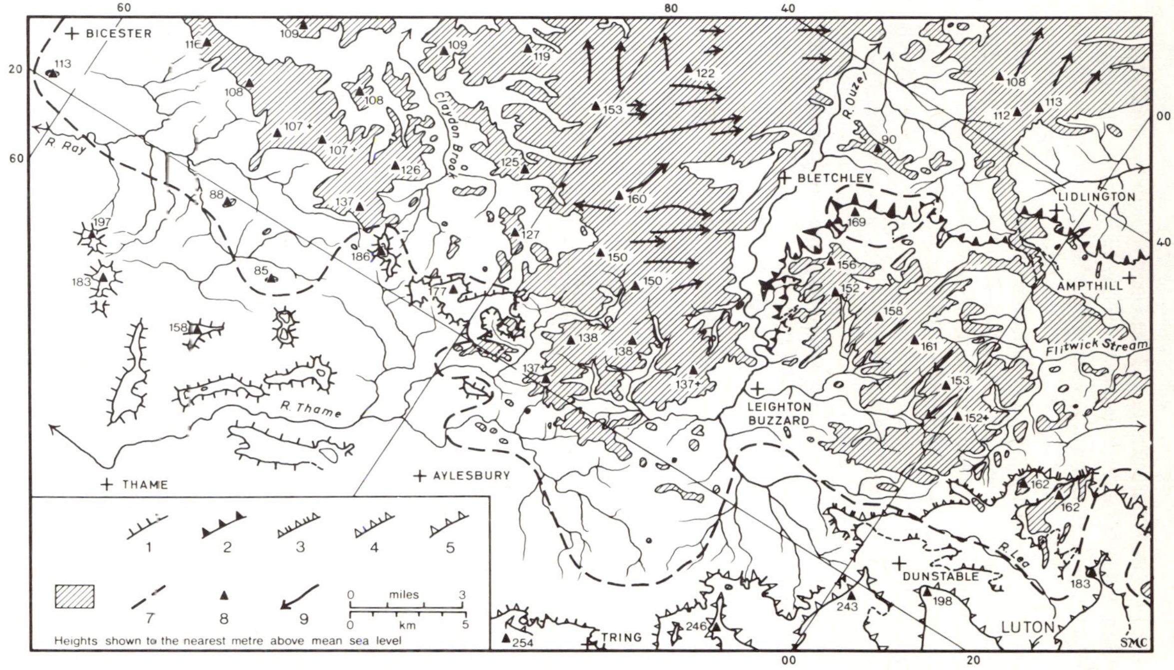

Fig. 12.7 The southern margin of the glacial drift in Oxfordshire.

The principal characteristics of the Gault vale to the southeast of the Lower Greensand cuesta are the many hills, giving a landscape atypical of the Gault. These hills, rising to over 150 m along the water parting at Toddington, northwest of Luton, are capped by glacial deposits. They appear to have been formed by the dissection of a drift sheet resting on the Gault and the Chalk Marl. They represent a late-mature stage in the dissection of the sheet of drift that covers so much of the East Midlands. These hill features continue west of the Ouzel where they gradually merge into the extensive drift plateau between Leighton Buzzard and Winslow.

The hills west and northwest of Aylesbury are rather higher and have steeper slopes than those capped by the boulder-clay. They have been formed in much the same way, by the dissection of a former plateau, but here the cap rocks are the Portland and Purbeck Limestones and the Shotover Sands. The hills to the north at Quainton and Oving formed embayments in the ice-sheet which projected between them in small lobes. The hills southwest of the former Great Central railway lay beyond the ice-limit. The present drift limit west of Quainton is an erosional feature, a till-capped bluff that can be traced through Edgcott, Poundon and Stratton Audley. There are a few small patches of boulder-clay around the plains of the river Ray, and these suggest that the maximum limit of advance of the ice-sheet reached to about the line of the A41 road. The most distant patch of drift caps Gravel Hill (113 m), south of Bicester. South of the drift edge both the Ray and the Thame have exposed extensive low-lying areas of Jurassic clay, although the Thame has been less successful owing to the lower elevation of the resistant rocks in its basin. The boulder-clay plateau north of the Ray basin is deeply broken by the Claydon Brook, a tributary of the river Ouse. The eastern tributary of this important stream is confined to the boulder-clay to the north of Quainton, but the western tributary rises well to the west of the glacial limit and only a little over a mile from the river Cherwell. It is an ancient stream of some importance.

The northern edge

The Jurassic plateaux lie between the river Welland and the river Wreak, which drains westward to the river Soar. The landscape here (Clayton 1959) is the most complex in the East Midlands, the main features are shown on the morphological map (fig. 12.8) and discussion must be confined to these. Oakham lies in the Vale of Catmose, a vale on the Upper Lias clays that is drained by the easterly-flowing Gwash and the river Wreak. The east side of the

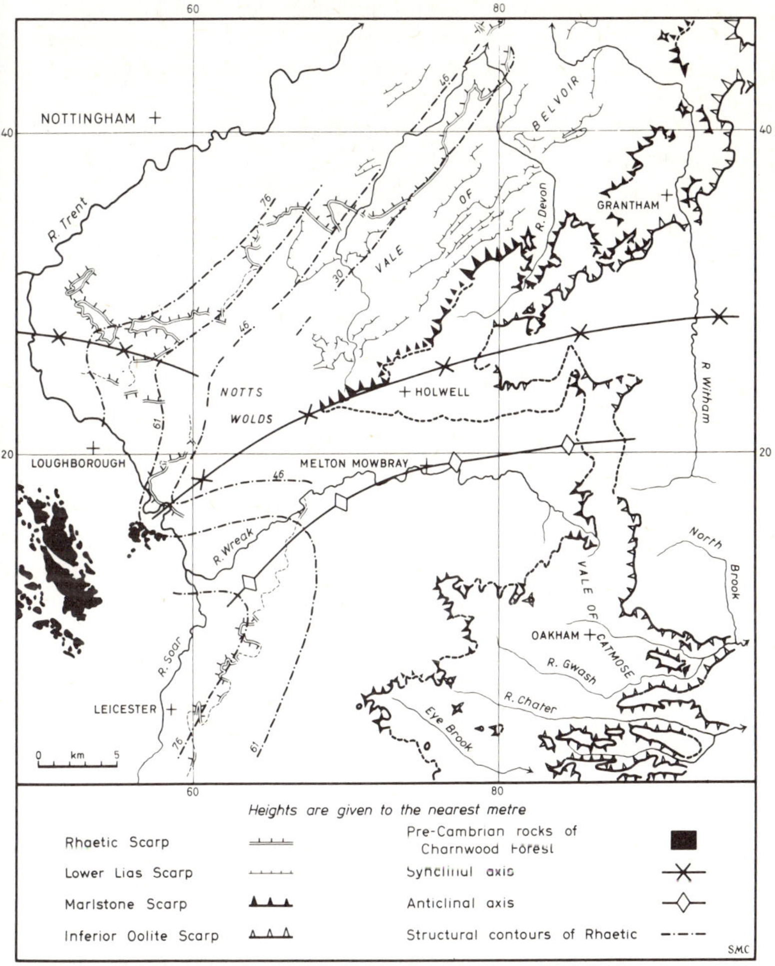

Fig. 12.8 Scarps and structures of the Melton Mowbray area.
This part of the Jurassic outcrop is marked by two pitching folds.

vale is the main scarp of the Interior Oolite rocks, and these also cap the ridges limiting the vale on the south. To the west rises the deeply-dissected backslope of Marlstone, with a series of outliers of Northampton Sand culminating in the Whatborough Hill (230 m). To the west of the Marlstone scarp a sheet of boulder-clay slopes gently westnorthwestwards towards the river Soar, dissected by parallel streams and showing a remarkably constant upper limit at about 167 m.

To the north of the Wreak the features are related to the Holwell syncline. The fold swings in a curve parallel to the Wreak, it can be traced to the Rhaetic outcrop at Barrow-on-Soar, to be lost in the Keuper Marl beyond. Eastward it dies out and does not affect the Great Oolite. The resulting westward swing of the outcrops is well seen in the three resistant beds crossed by the fold, the Lincolnshire Limestone, the Marlstone and the Rhaetic. The Marlstone here is almost at its maximum thickness, with the Ironstone it includes 14 m of resistant rock, and on the north side of the fold a high, northward-facing scarp overlooks the Vale of Belvoir. South of the axis the Marlstone makes no feature at all as it is buried beneath a thick sheet of boulder-clay which slopes quite steeply southwards towards the Wreak.

West of Old Dalby, at the western end of the Marlstone scarp, the plateau is continued, at a slightly lower elevation, by the boulder-clay forming the Nottinghamshire Wolds. The proven thickness of this drift is 22 m and the maximum thickness exceeds 30 m (Lamplugh *et al* 1909); as a result, although the streams are quite sharply incised, only a few have exposed the underlying rocks. The drainage pattern, particularly in the south, is markedly radial, flowing outwards from Six Hills (138 m O.D.). Between the Wolds and the Soar valley the Rhaetic forms tabular outliers; those around West Leake appear to be related to a second, minor syncline that may very well continue westwards to the Rhaetic outliers of Needwood Forest.

The Cheshire Plain

West of the Pennines and the interrupted outcrops of Palaeozoic rocks lying east of the Red Rock fault and its southwards continuation lies the Cheshire Plain. As a structurally-defined basin or cuvette this was well-established throughout Triassic times, and the lowland largely coincides with the Triassic outcrop between the

Pennines and North Wales. This lowland has been intensely glaciated, with a vigorous ice-flow southward out of the Irish Sea basin and a subsidiary addition of ice from North Wales along the western side. In earlier glaciations this ice reached the Bristol Channel area, and perhaps even into Avon, but it was the Devensian advance which completely filled the basin to dominate the landforms of the plain. As is common in lowland areas further north in the British Isles, the fresh landforms of deglaciation, although individually of modest dimensions, control the appearance of the landscape, the pattern of

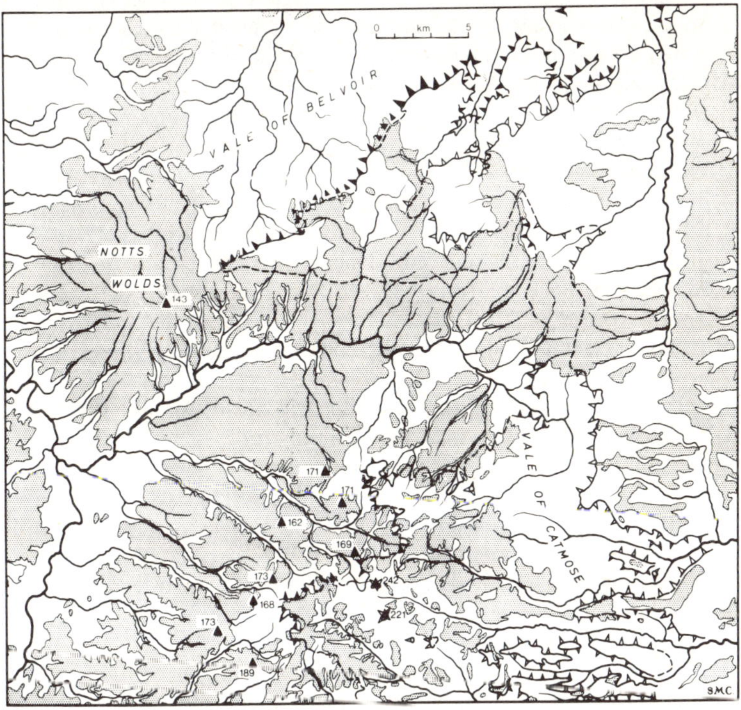

Fig. 12.9 The drift cover of the Melton Mowbray area.

The contract between the drift free Vale of Belvoir and the drift cover of the Marlstone outcrop and the Nottinghamshire Wolds to the northwest is seen. Another drift free area is the Upper Lias clay Vale of Catmose. The drift west of the Marlstone scarp south of the Wreak valley forms a pediment that slopes westward to the Soar.

slopes and valleys, and the characteristics of the soils.

Although the landforms of lowland deglaciation are deservedly emphasized in any account, the thick sheet of drift does not completely obliterate the relief related to the underlying Triassic. A broken north-south cuesta, best developed in Delamare Forest just south of the Mersey, marks the outcrop of the Keuper Sandstone within the thick Triassic succession. It reaches 143 m at Helsby Hill and 211 m at Bickerton Hill, 20 km to the south. The ridge is unlikely to have survived as an unmodified relic of a preglacial landscape, and its present form must reflect at least some differential erosion by the ice. This is supported by Gresswell's (1964) description of the three 'iceways' at the littoral margin of the plain which cut right through the sandstone outcrops of the Wirral and Liverpool areas. The same sandstone outcrop swings round the central basin (the highest beds, actually of Liassic age, occur near Wem). It is locally obscured by glacial drifts, perhaps after erosion by the passage of ice, and also rather broken by faulting. Nonetheless, it forms hills on the eastern side of the basin, as at Grinshill Hill and the hills between Wixhill and Hodnet, which reach 207 m in Hawkestone Park, halfway between Wem and Market Drayton. Apart from these hills, there are few other signs of the underlying Trias and the rest of this section will concentrate on the landscape left by deglaciation.

The literature on the Quaternary deposits of the Cheshire Plain is on the whole far more concerned with the stratigraphical succession and the age-relationship of the glacial deposits than with the surface forms. The persistent problems of interpretation arise from the misleading simplicity of a ubiquitous tripartite model of lower till, middle sands and upper till (Simpson 1959; Worsley 1967, 1970). While locally a tripartite sequence of this type is quite often present, there is no evidence either that the upper (or lower) tills are everywhere of the same age, or that the upper and lower tills at any one site represent successive (separate) glaciations. Further, the intervening sands are now understood to include two separate elements, the lower belonging to the Chelford Sands Formation (Simpson and West 1958) and dated around 60000 years B.P. (Brørup Interstadial), the upper forming part of the glacigenic Stockport Formation (Worsley 1967) and associated with the glaciation of the Cheshire Basin which occurred less than 30000 years B.P.

The maximum extent of the Devensian ice, when it passed beyond the Cheshire Plain to reach south of Wolverhampton and east of Stoke-on-Trent has been radiocarbon dated as later than 30650 years B.P. for it overlies peat of that age at Four Ashes, 8 km north of Wolverhampton (Morgan 1973). It might seem reasonable to assume that the progressive withdrawal of the ice from the Cheshire Plain led to a series of drift surfaces of progressively younger age. Some evidence of this was put forward by Boulton and Worsley (1965), who described a contrast in depth of leaching of calcium carbonate north and south of the Whitchurch moraine. Depth to calcium carbonate averaged 2.64 m south of the moraine, only 1.37 m to the north. Work elsewhere has, of course, established that time is not the only variable in such circumstances, but it is likely to be one of the factors in this particular case. Unfortunately the few radiocarbon dates so far obtained for other parts of the Cheshire Plain do not give us any idea of the rate of retreat, and it is possible that wide areas of the ice stagnated givng little contrast in the date at which the ice disappeared. A shell incorporated into the till at Sandiway, 20 km east of Chester, yielded a date of 27000 B.P., but this offers no hint of how much younger the till itself might be. Similarly a date of 10780 B.P. from sediments overlying the till at Wyrnbury is surely appreciably younger than the age of the underlying till.

This lack of any firm data on age for a major till sheet well within the range of radiocarbon dating is very unsatisfactory and reflects the continued paucity of 14C dates for the United Kingdom. As we have seen, the Whitchurch moraine is a major 'retreat' moraine (i.e. it is not a moraine built at the absolute limit of the Devensian advance) and the leaching data suggest that it is a halt stage separating drift sheets of somewhat contrasted age. Support for a change in the regime of the withdrawing ice-sheet at this time comes from the vigorous invasion of Welsh ice which seems to have been contemporaneous with the Whitchurch stage (Poole and Whiteman 1966). Near Ellesmere the contact zone between the vigorous Welsh ice and the less-active Irish Sea ice that by then occupied the northern half of the Cheshire Plain is well exposed. Gravel pits show that a fresh and irregular series of proglacial landforms is built on hills and coarse gravels that show very well-developed ice-push and ice-contact features. The so-called Ellesmere moraine can be traced to the Severn valley at Shrewsbury (Shaw 1972), but here it is more subdued as at this stage the Irish Sea ice had withdrawn clear of the

Severn drainage. In time we shall know the precise age of the absolute limit of the Irish Sea Devensian advance and the age of the Whitchurch-Ellesmere stage: it is idle to speculate here. However, as Shotton (1977) has noted, there is at present no evidence that requires the Cheshire Plain-Irish Sea ice-sheet to be as young as the 18000 yr advance known from Holderness.

Worsley (1970) and others have noted appreciable contrasts in the landforms north and south of the Whitchurch moraine. Over the central part of the Plain this seems real enough, but it is not a distinction that applies to the whole area. In particular, towards the southern limit fluvioglacial deposition took place on and within an area of stagnant ice occupying the Severn valley between Buildwas and Shrewsbury and extending south towards Church Stretton. The result is an area of fresh ice-contact relief, no doubt readily preserved on the permeable gravels, but clearly of Newer Drift age and style. In addition, the open water of the Shropshire meres suggests a recent origin since such shallow lakes do not survive for very long. Consequently it seems best to attribute the subdued till sheet of southern Cheshire to the style of ice dissolution, rather than to later modification.

The area north of the Whitchurch moraine is more diversified, and hummocky moraine ridges, meltwater channels and small eskers are quite common. Major outwash plains occur because the overall slope was northwards towards the ice and the meltwater was often ponded: a good example (although complicated by dead ice blocks and subsidence due to salt in the underlying Trias) occurs in the main area of Delamere Forest. The I.G.S. (Poole and Whiteman 1966) have described drumlin-like features in a 'morainic' belt from the Dee near Worthenbury, through Malpas and Cholmondeley Castle in the lee of the Bickerton and Peckforton Hills. They have also linked this belt with an 'upper boulder-clay' ice-margin which they regard as contemporaneous with the elusive Lake Lapworth, the highest level of which they place at 100 m.

The eastern slopes of the Welsh upland rise from the drift-covered lowland of the Cheshire Plain with the steeply-dipping Carboniferous rocks. The upland-lowland boundary is generally marked by the Cefn-y-fedw sandstone of Millstone Grit age, which frequently forms a crest at over 300 m, usually a little higher than the Carboniferous Limestone immediately to the west. Thus most of the Coal Measure outcrop falls within the Cheshire Plain lowland, lies below 150 m and is largely covered with glacial deposits. Locally ridges rise above the till plain where sandstone bands occur within the Coal Measure sequence (and they are few compared with the sequence east of the Pennines) as in the Summer Hill-Winter Hill ridge just northwest of Wrexham. From these ridges around 200 m the ground slopes eastwards to the low ground along the Dee and its tributaries. The regularity of slope seems to reflect a combination of underlying structural control with the eastward dip on the Carboniferous rocks, glacial erosion continuing the slope west of the Dee iceway (Gresswell 1964), perhaps finished off with a depositional slope on the till since it is typically 50 m thick.

13 Preglacial landform evolution

Preglacial denudation

D.L. Linton identified the Proto-Trent as an ancient river rising in the highest ground of North Wales and pursuing an easterly course past Llangollen, across the Midlands south of the Pennines, and out to sea somewhere near the Wash (Linton 1951a). This fitted his concept of a regional eastward component in the early drainage of the British Isles; he suggested that this developed on a chalk cover after it was domed along the line of the Irish Sea (Linton 1951b). The river must have been diverted by capture in the Cheshire Plain at quite an early time marking the first stage in the development of an Irish Sea drainage. While it must have crossed the Potteries area at least above the present level of 200 m or so, it is more likely that capture predated the 300 m surface and could have been long before that.

Any such river would have acquired tributaries by subsequent extension, and in trying to reconstruct these we meet several problems. First there is evidence that in the eastern part of the area a transgression to about 200 m has led to a later stage of superimposition with closely-spaced consequents (see the more detailed discussion in Eastern England). Second, glacial interference has been severe, and in the case of the preglacial Soar has only recently been fully recognized. Other major derangements may still remain to be identified. Finally, the oldest courses we can identify, those of the rivers in the southern Pennines, show conflicting features that are not easily resolved. In his original reconstruction Lindon (1951d) drew attention to the apparent superimposition of the Derwent across the pitching folds of Derbyshire and suggested that here was an ancient Trent tributary, superimposed on the Carboniferous rocks. In a later reconstruction (Linton 1956) he took the view that

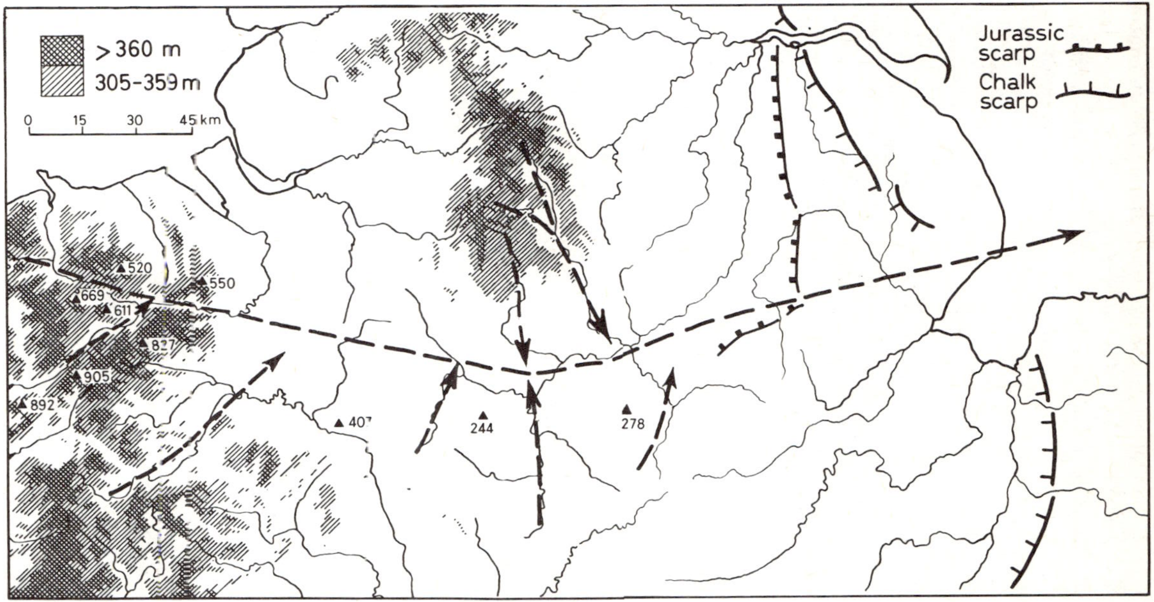

Fig. 13.1 The Proto-Trent of D.L. Linton

The Proto-Trent is one of the oldest features of the region, at least as old as the Summit surface of the Pennines, and probably originating at a higher level still.

the right-bank tributaries of the Derwent are older consequent streams, and that the Derwent has captured them as it developed as a subsequent stream at about the 300 m level, perhaps developing its present insequent stretches in the later process of cutting down into the Carboniferous strata.

The evidence is certainly conflicting, but there seems reason to follow Linton's later (1956) view. Although Pitty (1965) has suggested that some of the cols east of the Derwent may have been developed by down-dip breaching, the deeply incised course of the Wye across the Derbyshire dome is significant, while further north the Edale Stream is developed opposite a wind-gap at the head of the Rivelin valley at 360 m. Some of the drainage of the southwest Pennines, notably the Manifold, shows similar evidence of superimposition with partial integration by subsequents developed on the shales above the limestone. So far no attempt has been made to reconstruct the likely outcrop pattern of the upland plain above 300 m, but this seems the most likely stage for major subsequent development and this difficult exercise will need to be tackled. The importance of the former shale cover has been indicated by Warwick's investigation (1963) of the origin of the dry valley system on the limestone, and it seems likely to have played a very important role in the integration of the drainage.

The former right-bank tributaries of the Proto-Trent are now so broken up and disrupted by glaciation that their reconstruction is almost impossible. In preglacial times the Soar basin extended much further southwest than now, and the main water parting of England crossed from the Clent Hills to the northern Cotswolds near Moreton-in-the-Marsh (Shotton 1953). This course is closely parallel to the strike of the lower Jurassic and suggests that the Soar is a subsequent stream. While we might guess that it extended itself at the expense of the headstreams of such rivers as the Welland and the Nene, there is no prospect of reconstructing the drainage history of this low-lying, heavily glaciated area. The whole of the Midland drainage is now closely adjusted to structure; as we have seen each of the horsts is associated with a generally radial pattern of streams and these are gathered together by major rivers that occupy central positions in the intervening Mesozoic rocks. The circuitous drainage of the river Mease and the Anker with its tributary the Sence south of Ashby are good examples of this.

East of the Pennines the rivers show an approximation to a trellis pattern, reflecting the general scarpland character of the easterly-dipping rocks. On the Pennine flanks local relief is high and some remarkable adjustments of drainage occur. The best example is the precise inversion of the Ashover dome. Yet even here there are problems, for the westward turn of the Amber to join the Derwent cuts right across the structure of the Millstone Grit south of the dome (Clayton 1968). Further east the subparallel streams north of Nottingham have long been interpreted as superimposed consequents related to marine planation at about 180 m. Further north these relationships appear again with indications of adjustment to structure on the higher ground of the Millstone Grit and Coal Measures and of insequent drainage on the lower ground. Yet perversely the main stream of South Yorkshire, the Don, has deeply-incised meanders cut into the Grits southeast of Penistone where it is flowing southeastwards and then appears to be adjusted to structure as it flows northeast from Sheffield along the graben of the Don faults.

It is clear from this brief account that the evidence that can be obtained from the river pattern is very limited in character. The small area of high ground and the considerable interference of glaciation means that only a few drainage lines can be shown to be of any great age. The same is true of the evidence of the upland plains. The area above 450 m is so limited that little of value can be deduced in this particular area. In southwest Yorkshire and northeast Cheshire more than one author (Johnson and Rice 1961; Linton 1956; Sissons 1954) has identified an upland plain at about 500 m although there has been no agreement on its origin or on whether the rather flat summits at 600 m form a separate planation surface. Appreciable moorland surfaces transgressing both grits and shales suggest that this higher feature may be a relic of earlier planation rather than the adjustment of erosion to the upfolded and resistant grits at the base of the Millstone Grit series. Linton (1956) has demonstrated that in geometrical terms the observed dip of the Lincolnshire Chalk is adequate to carry it over the Pennines and has suggested the possibility that the highest summits might be close to a transgressive sub-Cenomanian plain.

The lower feature at about 300-400 m is far more widespread and we have some evidence of its age. In the south Pennines this is what Linton (1956) called the Upland Plain, and it forms a common skyline element in views in the southern Pennines. On the limestone where the upper parts of the valleys still lie at this level since they carry no streams, the surface is seen to have appreciable relief, certainly of the order of 100 m. But slopes are low and the contrast with the deep gorges (dales) that carry running water is clear enough. Around the rim of the upland the break of slope below this plain is commonly found at about 290 m, but this is not to say that it is present everywhere. However, this does seem to be the most common altitude where a break of slope occurs. South and southeast of the Pennines the highest ground reaches near this level: 315 m in the Clent Hills, 290 m in the Lickeys, 278 m on the isolated summit of Bardon Hill in Charnwood Forest, 319 m at the highest point of the northern Cotswolds. None of these relics is extensive enough to allow reconstruction of the surface or its connection with the Pliocene land surface of southeast England. Yet the connection seems indisputable and has been suggested many times. Over twenty years ago the author (Clayton 1953a) suggested that there was no lower surface in the Pennines that could be correlated with the Pliocene land surface of southeast England and conversely no lower surface in southeast England to correlate with this widespread surface cut on the resistant Carboniferous rocks.

As in so many of the British uplands the possibility of the exhumation of an upland from beneath Triassic rocks has received repeated attention. In the southern Pennines these arguments have been given added force by the occurrence of sedimentary beds in pockets in the limestone. That these are relics of a former cover is unquestioned, the problem is to identify the age of the beds and the height from which they have been let down. Are they in fact evidence of the age of the Upland Plain itself, or are they in fact let down from a former higher plain itself replaced by the Upland Plain? The lowest beds of these deep sinks may well come from such a higher surface. But the uppermost beds must surely be closely related to the present surroundings of these pipes. In places this includes glacial till from the last episode of deposition in the area and below this late-Tertiary fossil material is found, as at Brassington (Walsh *et al* 1972) where *Sequoia* trunks are known. These fills are the close analogues of the clay-with-flints of the high chalk of southeast England and they support a late-Tertiary (and probably Pliocene) date for the Upland Plain.

McArthur (1977) has attempted to date the surfaces in the upper

Fig. 13.2 The Upland Surface of the southern Pennines.

The Upland Surface of the Pennines is almost entirely restricted to the outcrop of the Millstone Grit and Carboniferous Limestone. It slopes gently southeastward, more probably reflecting the original eastward drainage (surviving in the Wye) rather than as the result of any movement along the western faults which lie well west of the highest summits.

Derwent basin by calculating the volume of rock excavated below the Summit Surface (*c* 600 m) and the Upland Surface (400-500 m). By assuming that excavation has proceeded at a rate of 50 mm/1000 yr, an elapsed time of half a million years is calculated from the initiation of incision below the Upland Surface. This would place it quite late in the Quaternary, but the surface itself may have been created and survived with little change for some time earlier in the Quaternary. Excavation below the Summit Surface is estimated to have taken 1-2 million years, and could therefore just be of pre-Quaternary age, but from this approach is more likely to be lower Quaternary. However, McArthur fails to allow for the long time with low rates of denudation required to form surfaces of low relief on such resistant rocks, and while the Summit Surface was close to local base-level, rate of erosion were presumably far lower than 50 mm/1000 yr (Young 1974). We may suppose that while it has only taken half a million years to destroy the Upland Surface, its initiation was about 1.5 million years ago, and initiation of the Summit Surface would then lie in the range 2-5 million years B.P. This is consistent with the stratigraphic evidence.

East of the Pennines several authors have recognized a closely-spaced stairway of planation stages. As Linton (1956) has argued, there is so far little sign that the various sequences proposed have any general validity, while the lack of recent work on the topic suggests little conviction that any attempt at regional synthesis would yield worthwhile results. Two critical problems emerge. The first is that the reality of the closer spacing of these features at the lower levels is out of proportion either to the amount of ground in which they might be preserved, or to the fact that their probably younger age would make them more likely to survive. There does seem to be a much closer spacing of these planation features at lower levels, to the extent that they can only be identified and mapped by careful fieldwork. This being the case they are necessarily identified in a subjective way, and in the absence of associated deposits it is hard to be sure that they result from successively-repeated episodes of planation. Their origin remains obscure, but they are significant features of the landscape and an element in the complex slope pattern to which soils and other apsects of the environment are related.

Most workers have concluded from the varying altitude of these features and from the way in which they are found within incised river valleys as valleyside steps, that they are subaerial in origin. They no doubt represent the accumulated effect of the many factors that can lead to a reduction of incision and the development of surfaces of low slope either as footslopes or as valley flats cut by river migration. Some are related to base-level movements, in many cases sea-level changes, in other cases to glacial ponding of rivers, river diversion by ice, and so on. Others will be related to the encountering of resistant beds as the streams cut down, although the existence of rock benches necessarily implies erosional stripping of unresistant rock above. Still others will be related to the changing regime of river flow with changes in climate, in vegetation cover and in the activity of slope erosion.

In an investigation of the eastern margin of the Pennines, Sissons (1954) suggested that most of the planation surfaces he was able to identify were marine in origin. As was argued in the discussion on his paper, most of the criteria he used are not diagnostic, yet one surface, the Bierley between 183 and 215 m, stands out as fundamentally distinct from the others. It is widely developed at the same height: analysis of steps on spur profiles based on the separation of the 50 foot contours of the one-inch map gave eighteen out of a possible twenty-seven for the 600-650-foot interval, the probability of this occurring by chance is only 0.000045. Two other factors (of the ten he suggests in his paper) point to a marine origin: the consistent height of the break of slope at the upper limit of the surface and the close spacing of subparallel consequent streams. As we have seen, this latter feature recurs in the Mansfield plateau north of Nottingham where such streams as the Meden, Maun and Rainworth Water flow east and dissect an extensive summit surface that crosses from the Magnesian Limestone on to the Coal Measures.

The low ground of the Trent valley precludes tracing the western limit of this feature unless it forms the higher part of the Keele surface (Yates 1956) in north Staffordshire. South of the Trent, glaciation has so far altered and concealed the preglacial landscape that evidence is again scanty. Kellaway and Taylor (1953), using subdrift data from areas of ironstone mining, postulated a tilted summit surface for the area of the Jurassic outcrop in Rutland and Northamptonshire. Their evidence for a single tilted surface is not strong, for the data do not allow precise alignment of the contours of the reconstructed surface. It has already been suggested (Clayton 1959) that in the Oakham area two distinct planation surfaces at

185-200 m (west of the Vale of Catmose) and at 149 m (very well-developed in the east, cutting across the Jurassic strata most convincingly) may be identified within the area included in the East Midlands erosion surface. It may well be that the 185-200 m surface is marine and associated with superimposed subparallel streams and that the adjustment to structure noted by Kellaway and Taylor is restricted to a lower area. In addition, they ignored the effects of glacial erosion, yet the till here contains much local detritus. To the south there is little ground as high as 185 m until we reached the Chilterns at Dunstable Downs where the early Pleistocene shoreline has been mapped at about 200 m (Wooldridge and Linton 1955). The gap of over 95 km precludes direct correlation of the two features.

Dury (1951b) suggested that a tilted summit surface of the same general form of the East Midlands erosion surface extended south-west from Northamptonshire into the Cotswolds. As this takes us beyond the glaciated area the evidence is less obscure than in the East Midlands, and in many ways the area provides a useful testing ground for such a concept. At least one unpublished thesis, found evidence for a stairway of features, rather than a single tilted surface, and although the evidence is purely morphological and unsupported by any deposits that can be dated, it is convincing. Here, as in the East Midlands, the concept of a tilted summit surface raises the problems of the time taken to form it and then to tilt it; and the evidence for absence of such tilting in surrounding areas means that the concept of a tilted surface requires stronger evidence than the rather crude maps of generalized contours so far produced.

14 Glacial and postglacial landform changes

Outline

Morphologically the most significant subdivision of the glacial deposits of Great Britain is into the Older and Newer Drift (Wright 1914). The limit of the Newer Drift deposited by the last ice-sheet lies along the western Pennines and to the east of the Cheshire Plain. In the east the margin was generally well away from the boundary of the area covered in this chapter as Straw described earlier in this book. Most of the lower ground was glaciated during the early glaciations, and in the east and southwest Midlands there are wide areas of till lying on the plateaux. In some areas the glacial deposits are so thick that they form a very different relief from that on the solid rocks they bury. While it might be thought that the maximum limit of glaciation would be well established, this is actually far from the case. Meltwater channels along the Pennine flanks have been shown to be unsatisfactory evidence for ice-front positions and the increasingly scattered nature of drift relics towards the glacial limits (particularly east of the Pennines) makes the location of the boundary very difficult. It is often suggested that a large part of the southern Pennines upland was not glaciated, while the Cotswolds and part of the Chilterns were more certainly ice-free. However, it should be noted that the general tendency of research is to identify new deposits of glacial drift and so reduce the areas thought to have escaped glaciation, while reconstruction of the maximum extent of the British ice-sheet makes it difficult to envisage ice-free uplands in the Pennines and elsewhere, at the maximum of the Anglian glaciation (Boulton *et al* 1977).

While the general utility of an Older Drift/Newer Drift division has often been demonstrated, any detailed consideration of the relationship between glacial deposition and river erosion will suggest further subdivisions. In some areas a very early glaciation can be detected, as in Wills' (1924) recognition of the 'First Welsh' (Anglian) glaciation. In other areas the main part of the Older Drift is found to consist of two separate tills, often with gravels between. It has often been argued, too, that the Newer Drift includes at least two separate advances; such arguments have been based either on the relationship of glacial advances to river terraces, or the relationship of the drifts to the present valley forms, or on the presence of stratigraphic sequences including both tills and gravels. Some of these ideas will be touched on in the following pages.

In many ways the different directions of ice movement are as important in the Midlands as the chronology of the glacial advances. The Pennines divided the northern ice-stream into two major lines of movement down the Lancashire and Cheshire Plain and along the lower ground to the east of the Magnesian Limestone. Each of these streams was reinforced: the western by ice from the Irish Sea basin and Wales, the eastern by ice from the North Sea. Converging south of the Pennines these two great ice-sheets vied with each other for territory. At times the western ice was most powerful, and this is certainly true at the maximum of the Devensian glaciation when it reached much further south than the eastern stream. But during Wolstonian times eastern ice forced southwards and westwards to leave flinty and chalky drifts over almost the whole of the Midlands and as far southwest as the Cotswolds. But even in these earlier glaciations there were periods when the western ice, by arriving first or retreating more slowly, extended into territory also covered by eastern drifts so that the superimposition of lithologically-distinct tills is quite a common occurrence in the Midlands and does not necessarily indicate two distinct glacial phases (Harmer 1928). The deposits of these two ice-sheets are normally readily distinguished, for although the western ice brought in some flint, presumably from the bed of the Irish Sea, it is far less common than in eastern drifts and normally lithologically distinct. A complication in the middle Trent valley is that some ice moved over the Derbyshire Pennines, introducing a third lithological facies in that area.

The unglaciated area

The absolute margin of glaciation may be defined on the basis of several criteria. One limit surrounds all areas of till, and in this case new exposures may identify new till patches and lead to an extension of the postulated ice margin. In some areas moraines, often complex associations of unstratified drift and stratified ice contact deposits such as kames, mark the limit of an advance. Particularly where the ice-sheet margin lay against rising ground, marginal or submarginal meltwater channels may be associated with the ice-limit. In this case it is clearly very important to establish whether the highest channels are marginal or subglacial; in one case they mark the ice-limit, in the other they are an unknown distance within the boundary.

Both types of evidence have been used to establish a glacial boundary of Devensian age along the west and southwest Pennines. The earliest comprehensive account was by Jowett and Charlesworth (1929), and they described a system of proglacial lakes west of the Pennines with meltwater (and an ice-lobe) escaping southeastwards into the Derwent valley through the Dove Holes col north of Buxton. As has since been pointed out by Dalton (1958) among others, their hypothesis of a largely unglaciated south Pennines rests on the lack of till exposures on the higher ground, while their argument confuses the deposits of more than one glaciation. The ice that passed through the Doveholes col and down the Derwent valley was probably of Wolstonian age as Straw and Lewis (1962) suggested from the relationship of the patches around Bakewell to the present Wye valley. In the absence of interglacial beds it might be older but is very unlikely to be younger. However, the ice-margin which Jowett and Charlesworth traced along the steep western flank of the Pennines is of Devensian age. Johnson (1965a) has related the drifts and associated meltwater channels to the glacial deposits of the Cheshire Plain. Whether this ice-margin has overrun and obscured the Wolstonian margin we cannot say. We can be sure that the Devensian lay against but not over the southern Pennines. We also know that an active lobe of ice moved through the Doveholes col into the Derwent valley, probably in Wolstonian time. Whether this was an isolated lobe and the rest of the southern Pennines were then ice-free is less certain. If there was ice, it was relatively clean and has left little drift behind.

As we have already noted, the ice-limit on the eastern side of the Pennines is quite undefined. The Wolstonian margin seems to have been clean ice, leaving little drift. If meltwater channels were formed, they must have been few in number and have since been integrated into the stream network, making them difficult to recognize. In terms of an explanatory description of the landforms this boundary is not important to us as there is no obvious contrast between ice-covered and ice-free areas. But it remains a major lacuna in our knowledge of the limits of the Wolstonian ice-sheet.

Further south there seem to be no isolated nunataks. In the past such areas as Charnwood Forest have been thought to have escaped the ice, but ample sections associated with pipelines, motorways and similar engineering works have established that the ice covered all the hills, perhaps to an appreciable depth. To that extent, the classic exhumed Triassic landscape is rather more polygenetic than Watts (1903) thought.

The absolute southern limit of glaciation in the British Isles curves north and south with the relief. The Cotswolds and the Chilterns extend northeast as unglaciated cuestas, while between them broad lobes extended south along the vales towards the line of the Thames. In the west the margin is not very clearly defined on the Cotswolds dipslope, although extensive sheets of flinty and chalky till occur quite near the ice-limit. In the clay vale between Oxford and the Chilterns the margin is in part lost by erosion, for the river valleys have been lowered since Anglian (?) times, and the till patches survive on plateaux, often capping hills cut on the Oxford or Gault Clays. The Lower Greensand at Woburn forms another unglaciated re-entrant, although further north the till is banked north of the former scarp and the line of the Lower Greensand scarp is continued in a bluff capped by till. South of Oxford, glacial drift is found south of the Thames, this is the so-called 'Northern drift' (Sandford 1929) and from the very great dissection of the drift sheet is generally regarded as Anglian in age. Certainly it is lithologically distinct from the flinty Wolstonian till, and it may even represent extreme southward penetration by Wills' 'First Welsh' ice-sheet.

The Older Drift (Anglian and Wolstonian glaciations)

In his work on the West Midlands, Wills (1924) recognized an early 'First Welsh' (Anglian) glaciation and later 'Second Welsh' (Wolstonian) glaciation. Contemporaneous with the latter was the 'Main Eastern' (Wolstonian) glaciation which brought flinty and chalky till from the chalk outcrops east of the Pennines. Essentially distinction between the two western glaciations rested on geomorphological evidence, for the deposits of the 'First Welsh' ice-sheet lie high (about 60 m) above the present valley floors. They are also related to an early pattern of drainage long since disrupted by glaciation. It is hardly surprising that these relatively high-level deposits are now poorly preserved for they have been subject to further glaciations and to a tremendous amount of later fluvial dissection. As is so often the case, the evidence is clearest where the ice extended further than later advances, and a lobe that crossed the Jurassic escarpment in the Moreton gap reached south towards Oxford. Its limit down the Severn valley was thought to lie (Wills 1937) near Gloucester with the western side of that lobe resting against the Malvern and Abberley Hills. More recently Hey (1958, 1959) examined the evidence here and suggested that the Woolridge terrace west of the Malverns may not mark the marginal drainage of this early stage. However, this does not necessarily affect the inferred location of the ice margin.

As Wills noted, eastern drift of Anglian age is very poorly represented in the Midlands. The local occurrences of a lower till in the East Midlands have often been assigned to this advance, and similarly Shotton (1953) has suggested that the Bubbenhall Clay which occupies the floor of the former Soar valley and is now buried beneath the deposits of Lake Harrison, is of the same age. The problem here is that these deposits are all related to the same landscape as that on which the Wolstonian deposits rest. This is particularly true of the Bubbenhall Clay which lies low in the landscape covered by the advancing Wolstonian ice-sheet. In contrast Wills' 'First Welsh' till of Anglian age is everywhere well above this land surface and had clearly suffered much dissection before the next glacial advance. That this morphological separation is significant is further suggested by the position and relation to the present landscape of the contemporaneous 'Chiltern Drift' (Wooldridge 1938) of the extreme southeastern corner of the region. Here is the morphological equivalent of the 'First Welsh' drifts of the West Midlands and the 'Northern Drift' of Oxford and a clear candidate for an Anglian age. The conclusion would seem to be that most of the lower tills of the eastern drift sequence are in fact an early phase of the Wolstonian advance. This is not to say that the time-break indicated by the morpholigical distinction we have made is an interglacial period, and it may be that the warm interglacial break recorded by fossil material lies between the two tills of the Wolstonian advance. Geomorphologically, however, that particular break is not of major significance. The Midland and East Anglian correlation is discussed by Shotton, Banham and Bishop (1977).

The scattered nature of the evidence of the Anglian glaciation makes any general summary difficult. Only in the West Midlands is any general account possible and this is still almost completely dependent on the work of Wills, although allowance must also be made for the changes in the relief of the area during the Wolstonian glaciation which have since been described by Shotton (1953).

The retreat of the Anglian (First Welsh) ice-sheet (and its eastern counterpart) was followed by a period of erosion. Terraces were no doubt deposited as in the Thames valley, but with the possible exception of the Woolridge terrace, they have not survived further north. The rather small lower valley of the Stour has been so transformed by the development of the Avon and the later Severn diversion through the Ironbridge gorge that no recognizable deposits of this stage survive. By the close of the period the Soar flowed northwards to pass through Leicester below 60 m O.D., so that it seems likely that by this stage the Trent had been diverted from the Ancaster Gap for the main floor there preserved as flats on the spurs is at nearly 60 m and is in any case regarded by Straw as pre-Anglian. In the upper Thames and Evenlode valleys the Combe and Freeland terraces probably mark the valley floor at about the time of the Anglian 'Northern Drift' (Arkell 1947). These lie about 55 m above the present floodplain and antedate the Hanborough terrace which is well developed on the Cherwell at about 43 m above the present floodplain. Bishop (1957) correlated the Hanborough terrace with the Hoxnian interglacial noting that this date seemed to be confirmed by the Early Acheulean or Late Abbeville (Chellean) implement found at Long Hanborough. However, the sedimentology, fauna and included periglacial structures show that it is Wolstonian (Briggs and Gilbertson 1973).

The (double) Wolstonian advance of ice into the Midlands is a very important phase in the development of the area, for the till forming much of the present surface was deposited at this time. In addition, Shotton (1953) has shown that the shift of the main water-parting from a position near Stratford to its present line east of Coventry occurred during this period. His work on Lake Harrison has allowed the reconstruction of a pattern of events which is far more detailed than that known for any other area. The thick series of deposits that filled Lake Harrison and which is still preserved in the upper Avon valley naturally provides a far more complete record of events than either deposits of till or the scanty record of erosional development beyond the glacial limit. While these deposits are not particularly well exposed in quarries (and not exposed at all in natural sections) Shotton has used augering to establish the sequence and the lateral variation. Variations in lithology have been related to

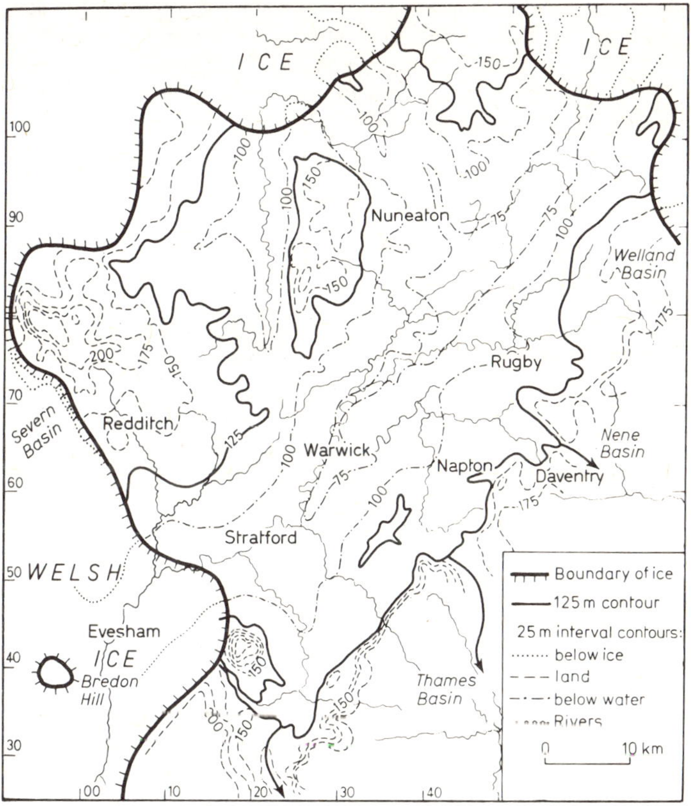

Fig. 14.1 Glacial Lake Harrison.

The contrours on the floor of Glacial Lake Harrison are broken. The margin of the lake may not have been occupied all at the same time, as the map represents its maximum extent.

environmental changes resulting from fluctuations in the ice margin or changes in lake level as outlet channels altered.

The plotting of subdrift altitudes from surface exposures and borings shows that the present Avon valley occupies a northeast-trending valley that slopes from a little below 90 m between Evesham and Stratford-on-Avon to 75 m at Warwick, and 60 m south of Leicester. This valley is as much as 8 km wide (measured between contours 15 m above its floor) and must represent the valley of the Soar before the advance of the Wolstonian ice. The former water-parting (the main divide of England) at its head has been so completely destroyed by later erosion that it is difficult to reconstruct, but it cannot have been very high and no doubt resembled the low divide between the Fens and Thames drainage below the Chiltern scarp. This was the Lias Clay vale at a time when it still existed as a continuous physiographic feature.

As the eastern ice advanced southwards up the Soar valley it blocked the drainage and formed a proglacial lake. At first this may have spilled over the main divide to the Bristol Channel, but this was not long-lived for western ice advancing down the western side of the divide reached the Jurassic scarp and forced the lake to rise, until it could spill over one of the Cotswold gaps to drain to the Thames. At this stage the lake had a surface level of about 125 m and any cols below this height southeast of Leicester or west of Stratford-on-Avon must have been blocked by ice. The lake must have extended into the Market Bosworth area and into the Tame valley, isolating the Warwickshire coalfield as an island. There are three cols across the Jurassic escarpment at about the height of the lake surface; the Daventry Gap, Fenny Compton Gap and the Dassett Gap. Bishop (1957) has shown that the Fenny Compton Gap (to the Cherwell) was the major outlet, although he was not able to rule out the operation of the others at some time or other, and in addition showed that the Moreton Gap functioned when the ice advanced furthest south. Dury (1951a) mapped wave-cut benches along the lake margin in southeast Warwickshire.

The maximum extent of the lake did not last long, for the ice continued to advance, reducing the lake area and at the same time filling it in with vast quantities of sediment. Below the lake sediments is the Bubbenhall Clay — small patches of till that probably represent earlier deposits in the valley, no doubt when the low divide to the Bristol Channel was not blocked. Above this lies the Baginton-

Lillington Gravel and the Baginton Sand, deposits of streams draining the higher ground to the south, and representing in large part the accelerated erosion of glacial drifts related to periglacial conditions near the ice-sheet. Shotton regarded these deposits as the indicators of the advancing ice, but two points suggest a slightly different interpretation. In the first place, the gravels are strongly related to the dissection of drift deposits and most of the stone consists of erratic Bunter pebble or flints. Second, considerable erosion of these beds suggests not simply periglacial conditions but the type of instability associated with the lack of vegetation cover. Can these deposits represent the retreat of ice (perhaps that which deposited the Bubbenhall Clay) that later readvanced to impound Lake Harrison? While Shotton regards the Bubbenhall Clay as probably of Anglian age, we have already noted that it valley-bottom position (in relation to the landscape into which the Wolstonian ice advanced) suggests that it may be younger.

Whatever the details of this first stage, a proglacial lake was soon established and the Baginton Sand is overlain by the Thrussington Till (Shotton 1977) which ponded the lake and the main lake deposits, the Bosworth Clay and Silt and the Wolston Sand, and above these the Oadby Till. The Bosworth Clay and Silt consists of up to 18 m of laminated lake clays (the lower part of the latter frequently almost indistinguishable from the Keuper Marl although scattered pebbles were dropped from floating ice). As the period of deposition of the Bosworth Clay proceeded it became more calcareous and paler in colour. The Wolston Sand covers a huge area (650 square km) and Shotton regarded it as marking the advance of the Oadby Till ice. Shotton has also been able to demonstrate from the sequence of arrival of rafted erratics in the lower part of the Bosworth Clay, how predominantly Northern Drift carrying Bunter material was displaced by eastern drift bringing flints, and also the relatively rapid advance past the inliers of igneous rocks outcropping south of Charnwood Forest. An early advance past Croft was followed by retreat to allow deposition of lake deposits to the north of Leicester.

Lake Harrison came to an end not when the eastern ice retreated (although that had started), but by the much more catastrophic retreat of the other impounding ice-sheet in the west. This must have exposed an appreciable thickness of lake deposits that had been banked up against the ice, and the infant Avon rapidly cut back northwestwards. Reinforced by the meltwater still moving into the former lake basin from the easter ice front, headward erosion was extremely rapid and by the time the Soar could re-establish itself with the withdrawal of the ice north of Loughborough, the divide on the lake fill had been pushed northwards to very near its present position. It is interesting to note that the retreating eastern ice-sheet left deposits of outwash gravel (the Dunsmore gravel) over a wide area, but that tills associated with the advance are hardly ever found except further north as around Leicester. It would seem that the ice retreat left unvegetated moraine which was rapidly resorted by the meltwater. No doubt too the unvegetated character of the newly-exposed Lake Harrision floor was a further factor in the rapid evolution of the river Avon.

Nowhere else are the events associated with the Wolstonian glaciation so well documented. North of the Lickey and Clent Hills small proglacial lakes were impounded during the advance of the ice, lake deposits are found in many temporary sections in parts of Birmingham, and have been described by Pickering (1957). He showed how the Rea valley earlier lay further west, so that on the western slope of the present valley two lacustrine levels occur beneath the main sheet of till. These two lake levels represent proglacial damming of this early Rea valley. A lower lake level in south Birmingham is an arm of Lake Harrison.

The sequence of events in the Trent valley has already been discussed by Straw in the section on Eastern England, and will not be repeated. However, two other aspects of the pre-Devensian glaciations warrant discussion here: the evidence for multiple glaciation in the East Midlands, and the related problem of the varying extent of the eastern, western and northern (Pennine) ice-sheets in the Midlands. It is so far not possible to give anything like a complete account of the complex history of the divide between western and eastern ice in the Midlands. Further, Shotton's work on the early stages of Lake Harrison shows that even with a full sedimentary record interpretation is difficult. In the Wolstonian glaciation northern and western ice entered the area first, and it was at a later stage that the eastern ice gained in power and swept far westwards in the Middle Trent valley and southwestwards into Lake Harrison. This sequence of events was also described by Rice (1968) in central Leicestershire. Flinty or chalky drift is generally the upper deposit in the Midlands, and while many sections near Nottingham and in the Jurassic scarplands show chalky Jurassic boulder-clay, the underlying gravels are full of Carboniferous and Bunter erratics. This overlap of facies has led to the interpretation of the lower deposit as the remnant of an earlier glacial rather than an earlier phase of the same glaciation. In the absence of interglacial deposits this problem must remain unsolved, although it is possible that a morphological investigation of the occurrences would help to establish the extent of erosion between the deposits of the Northern and the Eastern Drift.

The problem applies in particular to the age of the till found on the limestone plateau of south Derbyshire. The ice that entered the Soar valley in the first stages of Lake Harrison was 'northern' in character and had probably arrived down the Derwent valley. Till found by Waters and Johnson (1958) in the central Derwent valley predates four valley stages or terraces, the highest of which (the Hathersage) is about 50 m above the present valley floor. This till is thus likely to be of Wolstonian age, and we have seen that the ice came over the Doveholes col. More widely scattered patches of drift on the plateau top might be related to the same glaciation, but may well be related to an earlier Anglian phase. Pennine gravel found east of the Trent at Wilford Hill is now at about 75 m (or 50 m above the floodplain), and from our knowledge of the Soar valley is likely to predate the initiation of Lake Harrison. Alternatively it represents the Wolstonian advance, but was deposited on the valley slopes. Isolated patches of fluvioglacial gravel found on high ground northwest of Derby (e.g. at Blackwall) are very reminiscent of Wills' descriptions of First Welsh deposits in the west Midlands and are very likely to be of Anglian age.

Bernard Smith summed up the Trent valley south of the Pennines as 'a cockpit where tongues of ice from various directions strove for territorial rights' (quoted in Swinnerton 1937). Thus we have already noted patches of Pennine drift in this area. A Lake District erratic at Waterloo Clump, across the river from Burton-on-Trent, must mark an incursion of western ice of Wolstonian age (although taken to be Devensian by Wills 1950, fig. 29) as does the huge boulder of Denbighshire grit in the grounds of Birmingham University. Later the eastern ice became dominant, as to the south, and flinty till was carried west to Uttoxeter and flinty outwash poured down the Blithe valley west of Abbots Bromley. The curved northern boundary of this lobe shows no sign of deflection by Pennine ice so that by this stage it was either held back by the eastern ice or had stagnated due to a cessation of flow across the Doveholes col. The power of eastern ice during these later stages of the Wolstonian glaciation does suggest a westward drift of the main flow (probably due to increased flow through the Wash gap as it was scoured and deepened) and this may have led to glacial erosion east of the Pennines. Indeed, it may be that erosion was dominant here, accounting for the very small patches of till found across the outcrops of the Coal Measures and Magnesian Limestone (see also the account by Straw, p.55). It is very probable that glacial erosion formed the depression north of Gainsborough now filled by glacial and alluvial drift, and it seems certain that ice excavated the Vale of Belvoir. Here the contrast between the broad, almost featureless vale with its very thin and scattered patches of till and the extensive till sheet south of the Marlstone escarpment is most readily explained by a change in the dynamics of the ice-sheet after it surmounted the Marlstone escarpment (Boulton *et al* 1977). To the west of the Vale of Belvoir the power of the ice declined so that the Rhaetic hills southeast of Nottingham survived.

The Newer Drift (Devensian glacial deposits)

Using the term as a time name, Shotton and others have described river terraces as part of the Newer Drift sequences. Here we restrict the term to the original sense of Jukes-Browne (1887) and Wright (1914) and will discuss only the glacial deposits (including direct outwash) of the Devensian glaciation. At this stage the eastern ice was far less powerful than that in the west and as Straw describes, it reached into the Vale of York and down the east coast to the Stickney-Sibsey ridge in Lincolnshire and Hunstanton in Norfolk. Periglacial conditions affected the East Midlands with many indications of deep permafrost. The more powerful western ice advanced into, and at places just beyond, the Cheshire Basin, bringing a highly distinctive till known as the Irish Sea drift. This readily-identified drift is well known in the Severn valley, for the first terrace of the diverted Severn below the Ironbridge spillway is the Main terrace, with a distinctive collection of erratic rocks (Wills 1938).

Despite the great contrasts between the Older and Newer Drifts, so ably summarized by Jukes-Browne when he first described them, it has proved very difficult to establish an acceptable limit to the Devensian glaciation. In part this is due to lack of detailed field mapping: most of the ground has not been tackled this century despite its interest and importance, while the published map of the Geological Survey has done little more than separate stratified from unstratified drift. A second factor is the presence of subdivisions within the Newer Drift itself. The Ellesmere moraine is a far younger and fresher feature than the drift nearer to the limit, while beyond the boundary of the Newer Drift occur little-dissected spreads that morphologically seem younger than Wolstonian drifts. Thus outwash from the youngest ice advance west of Abbots Bromley and Rugeley forms a feature in the Trent valley that is apparently younger than the Ipswichian (Beeston) terrace.

These drifts may represent an early ice advance of the Devensian Glaciation. Now that radiocarbon dating allows finite dates for the later part of the Devensian glaciation, some problems have been settled, and, as is so often the case, other difficulties have arisen. An important discovery was a buried peat from Four Ashes north of Wolverhampton, which gave a date of 30655 years (Shotton 1967). The till which overlies this peat is the Irish Sea till of Devensian age which extended eastwards to Wolverhampton and Cannock. Basal deposits from the kettle holes in the till sheet a few kilometres further north gave dates of 11660 and 11580 years (Morgan 1973), so ice advance is bracketed by these finite dates, and the maximum is likely to have been about 18-25000 years ago. A further date is of the Upton Warren interstadial, the site is a terrace of the Salwarpe southwest of Birmingham, and has been dated at 41500 years and 41900 years B.P. The environment was treeless and cold (Coope, Shotton and Strachan 1961).

The Upton Warren terrace may be traced down the Salwarpe to its junction with the Severn where it grades into the Main terrace of the Severn. The Main terrace is the major depositional evidence of the diversion of the Severn through the Ironbridge gorge, and thus represents the retreat stage of the Irish Sea-Shropshire lowlands ice sheet that caused the Ironbridge diversion. Further north the date on the Chelford deposits shows that the northern part of the Cheshire Plain was ice-free at about 60000 B.P. There is something of a conflict between the Upton Warren date and the Four Ashes age for the Devensian limit, and this has received a good deal of attention, but so far no agreed solution has been suggested. The most recent review is by Shotton, Banham and Bishop (1977).

The Ironbridge gorge marks the diversion of the Severn from a route northwards to the Dee to its present course. In general this diversion was brought about by the Irish Sea ice advance which filled the Cheshire Plain and so blocked the northward route. The col south of the Wrekin and north of Wenlock Edge must have been above 160 m, and was probably close to 200 m. This is well above

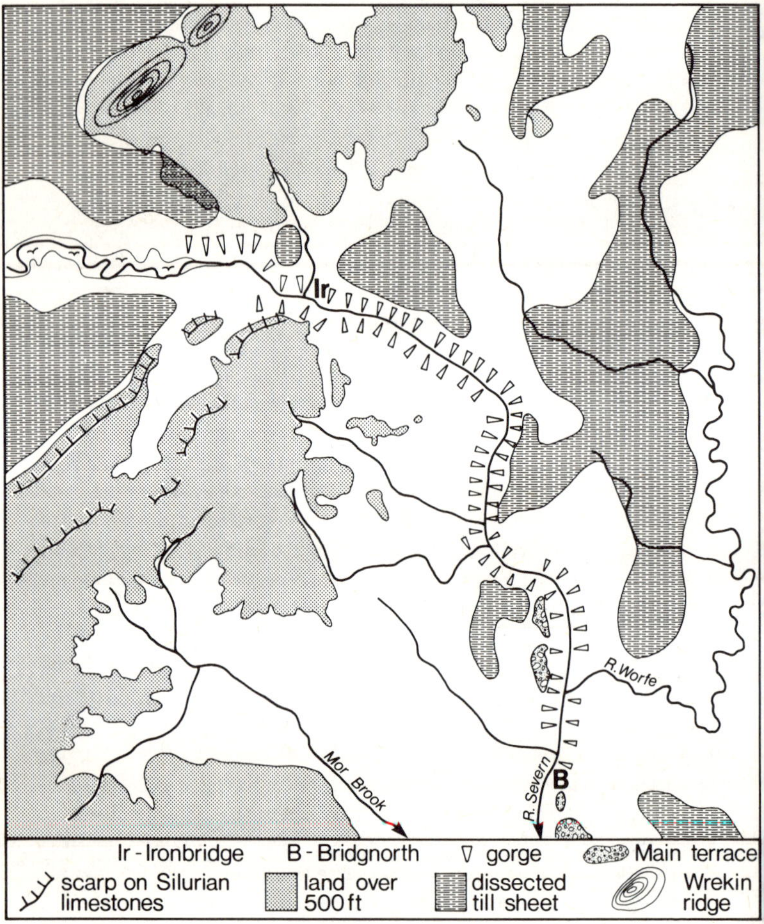

Fig. 14.2 The Ironbridge gorge.

This demonstrates how the gorge seems to have been initiated in an ice-marginal position.

160 m, and was probably close to 200 m. This is well above the existing height of the interfluve at the head of the Worfe basin northeast of the Wrekin where cols occur as low as 143 m. Thus there can be no question of the initiation of the course as a simple overflow from an extensive proglacial Lake Lapworth. The channel must have been started in an ice-marginal (and perhaps in part subglacial) position along the southwestern margin of the maximum extension of the Irish Sea ice-sheet. Incision would have been rapid enough for the course to be maintained when glacial retreat did occur. The new course carried the Severn drainage into a south-eastward flowing tributary of the Worfe system, and the new master-stream found itself with the Worfe, Stour and Avon as tributaries. The maximum discharge at the time of diversion is marked by the Main terrace, full of Irish Sea erratics, some being very large (up to 10 tonnes). Near Bridgnorth this is at 80 m and closely associated with flow till, so that by the time the terrace was formed the Ironbridge gorge must have been well-established and yet ice was still in the vicinity. The floodplain above the gorge is now at 45 m, so that incision has continued to be rapid to the present day. R. Hamblin of the Institute of Geological Sciences has recently mapped the sheet including Ironbridge. He reports a buried gorge of equivalent dimensions to the Ironbridge gorge, now full of sands, gravels and till, and clearly overridden by the Devensian ice. He has suggested it represents an overflow connected with an Early Devensian glaciation.

We have already mentioned the morainic features of the Cheshire Plain. These have been variously interpreted as marking significant time differences — major pauses in retreat — or simply marking phases when marginal forms were created that have survived. The most vigorous features are associated with incursions of Welsh ice into the plain, as at Ellesmere, although this is now interpreted by Francis as due to settlement over buried ice, not due to shearing as suggested previously. The ice seems to have been present at the same time as the main Irish Sea ice to the east. North of Wrexham a huge and complex outwash fan also included blocks of ice that have since formed kettles. It slopes down quite steeply on the east, but is no longer regarded as a delta into Lake Lapworth.

Following Johnson (1965a) the Devensian boundary against the western Pennines has already been described in outline. The complex oscillations of the glacier margin postulated by such earlier writers as Jowett and Charlesworth are no longer required by an hypothesis that includes subglacial channels. Indeed, it is now no longer easy to establish any clear sequence of events when so many features are attributed to water under pressure beneath the ice. Johnson was only able to map one Devensian ice-limit along the western Pennines, with the upper limit of erratics lying near to 365 m for about 55 km from Rochdale to Macclesfield. The same ice-sheet extended south-east through the Aire gap, penetrating the higher valleys around Bradford. Here again the published account (Stephens *et al* 1953) describes a complex arrangement of proglacial lakes connected by spillways, but it may be that much of the drainage was subglacial. The Aire ice was much less powerful than the Cheshire Basin ice, and south of Bradford on the Aire-Calder water-parting the till limit is at only 185 m and falls to about 120 m south of Leeds. By the time the ice lay at the Escrick moraine (which is not the maximum limit of Devensian drift) most of Airedale must have been ice-free.

Whatever the role of subglacial meltwater in cutting channels near the ice-margin, the meltwater was eventually carried away from the glaciated areas by subaerial spillways. We have noted the Ironbridge gorge which carried Welsh water south to the Bristol Channel. In the Pennines two great spillways carried water eastwards, the Todmorden channel in the north and the Churnet valley in the south (King 1960). The original intake of the Todmorden channel is now completely obscured by landslips that followed the intense incision. The water flowed eastwards down the Calder. The Churnet heads in an area of considerable alluviation, and remarkably confused drainage, at the head of the Trent. Its relations to the Dove terraces confirms that it carried water in the Devensian Glaciation although it could well have existed in an earlier glaciation. Some authors (notably Jowett and Charlesworth, 1929) have suggested that meltwater also escaped eastwards across the Doveholes col. The evidence is now obscured by quarrying, but this high col cannot have carried water for very long and has not been altered in the same way as the two other channels.

Shotton (1953) has convincingly demonstrated the sedimentary evidence for Lake Harrison, while Dury (1951a) had earlier identified its shore platforms in the field. It remains to comment on our knowledge of two other supposed proglacial lakes, Lapworth and Humber. Lake Lapworth was postulated by Wills as the inevitable result of the southward advance of Irish Sea ice into the Cheshire-Shropshire Plain, and the necessary precursor to the Ironbridge spillway. No convincing deposits have been found, yet we seem reluctant to abandon the concept of a large proglacial lake. A review by Worsley (1975) concluded the matter was still open; there are lake deposits in different areas, but none need indicate more than local ponding. Poole and Whiteman (1961) seemed in no doubt about the lake, and even referred to water stretching as far as Wigan and Bolton in Lancashire. Lake Humber is a much more extraordinary phenomenon. At its maximum postulated extent it would have submerged not merely the southern part of the Vale of York, and the lower Trent valley, but also extended through the Lincoln gap to cover the whole of Fenland to overflow across the low divide to the Waveney on the Norfolk-Suffolk border. The Fenland record is concealed by later sedimentation, and most of the evidence comes from the Vale of York. Here lacustrine stages are found, but opinions differ about the reality of the shorelines. The issue is fully discusssed by Straw earlier in this book.

Frozen ground

A large part of extraglacial Central and Eastern England was underlain by permafrost during the Devensian glaciation. In the most southerly parts of the region this can rarely be dated and some features may date from earlier glaciations when it was also ice-free. In the East Midlands the features can be shown to postdate the Older Drifts, while in Warwickshire huge polygons that almost certainly indicate true permafrost occur on the Avon terrace number 4 at Abbots Salford, north of Evesham, and are consequently younger in age. Shotton (1960) suggested these must have formed at the Irish Sea maximum. At one site in southwest Yorkshire, Shotton and Wilcockson (1951) were able to date the cambering to the Devensian glaciation. Other so-called periglacial disturbances are ubiquitous and are also found north of the Devensian limit. Indeed such disturbances as minor frost-wedge, casts, involutions and spreads of soliflucted angular gravel, are almost always met in sections, but as they do not involve the presence of permafrost, they do not necessarily represent a major cold period and convolutions can probably form today on the highest Pennine plateaux. Their chronological significance is rather limited.

The fact that most cases of cambering have been described from the East Midlands is simply a reflection of exploration for, and exploitation of, iron ore (Hollingworth and Kellaway 1944). The comparison of ridge crest and ridge margin levels for a sandstone layer is usually impossible without borehole evidence and though we may suspect cambering elsewhere it is not easy to prove. Similarly, large polygonal ice wedges, although occasionally seen on aerial photographs, are most readily identified in quarry faces and in areas stripped of their topsoil. The gulls associated with cambering can rarely be detected by normal field mapping.

Much current geomorphological literature emphasizes the speeding up of erosion under periglacial conditions, although Bird (1967) has shown that it need not be the dominant style of landscape reduction. Some deposits in this area show the effects of temporary increases in erosive action either during a cold period, or perhaps, more probably, at its onset or end. The fans of gravel below the Cotswold (Tomlinson 1941) and Chiltern scarps (Arkell 1947) are presumably broadly contemporaneous with the southward advance of the Older Drift ice sheets and represent deposits of a periglacial environment in the precise sense. Unfortunately their relationship to any flights of river terraces is obscure. Valleyside fans are known in the Cotswolds and

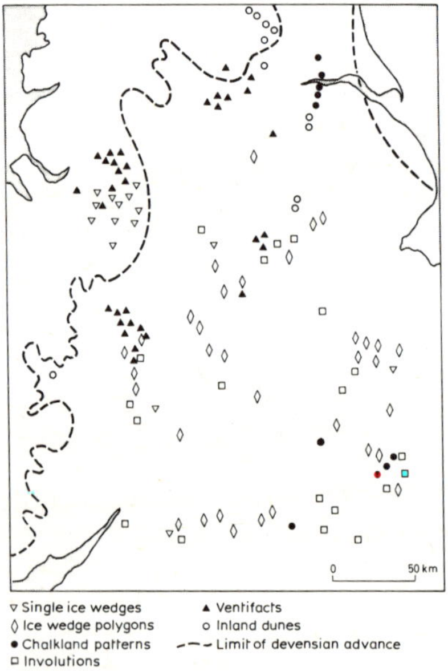

Fig. 14.3 Periglacial features in the Midlands.

the Pennines, while sections in solifluction detritus are common. No studies in this area have matched the important work of Ollier and Thomasson (1957) in the Chilterns on the development of valley asymmetry due to microclimatic differences on the two slopes. Waters and Johnson (1958) described four terrace stages of the river Derwent, one glacial phase, and a subsequent much younger periglacial phase during which the till moved downslope and covered the river gravels. The till was dated as Older Drift (Wolstonian or Anglian) and the periglacial phase was thought to be Devensian.

High pore water pressures associated with frozen ground have been related to slope instability in this area as elsewhere in the United Kingdom. Chandler (1971) described features on Lias slopes of the Welland valley near Uppingham related to periglacial conditions, and in another paper (Chandler 1970) related slides on the steeper southern slopes of small valleys to asymmetry which he explained as the result of more rapid degradation of the south-facing slopes under a periglacial climate. It is often suggested that the major landslides of the Pennine shale slopes capped by grits were most active in a past periglacial climate, but it is clear that even if movement was initiated in such conditions, it has continued into the Holocene. Thus Franks and Johnson (1964) described slides dated by palynology to zones VI-VIII (7200-2450 B.P.) at Charlesworth, while many of the large slide areas are still active today. Similarly, the movements described by Chandler (1971) were renewed when pore water pressures increased following Iron Age deforestation, and again since the seventeenth century, although they are apparently inactive today. On the other hand slides on the slopes of a meltwater channel at Walton Wood on the M6 were initiated by incision, not by climate, and were not fully stable when they were dramatically reactivated by engineering work (Early and Skempton 1972).

River terraces and terrace chronology

All the larger rivers of the area have flights of terraces associated with their present courses. These dissected spreads of occasionally fossiliferous gravels mark the adjustments of the rivers to changing discharge of water and sediment and to the fluctuating base levels provided by the glacio-eustatic shifts of Pleistocene sea-level or the more local effects of glacial ponding or diversion. All the major rivers flow out of this area to the sea, the Aire and Trent to the

Humber, the Thames to the North Sea, and the Severn to the Bristol Channel. The Trent has been so disturbed by the southward advance of ice-sheets that no simple terrace flight can be traced anywhere near the sea. The author (Clayton 1953b) has suggested that the terraces west of Lincoln were related to sea levels, but Straw (1963) believes the lower two terraces reflect ponded proglacial lake levels. The Severn flows into a long embayment of the Irish Sea with a very high tidal range and despite several attempts to relate particular terraces to base level (e.g. Beckinsale and Richardson 1964) none can be regarded with confidence. The Thames flows into a subsiding zone at the southern end of the North Sea, while in any case the links along its course from below London to the Colne area and through the Goring gap are very tenuous.

As Clayton has shown (1977), all geomorphological work on river terraces rests upon some important assumptions. In the first place the longitudinal continuity of the remnants is assumed and on this basis gradients may be calculated. In fact, with a few obvious exceptions, such as the Severn below Bridgnorth, these gradients are generally very close to those of the present floodplain as is recognized by such generalizations as referring to the Boyn Hill as the 100-foot terrace of the Thames. Significant departures are usually limited to upstream areas where the terrace may eventually pass into the modern floodplain at a knickpoint, and estuarine sectors where the modern river may be tidal and so quasi-horizontal while the terraces may continue to slope seaward, so coming closer to the river. The usual interpretation in such cases is to assume that they represent a river slope to a more distant coastline.

Second, the regime of the river is now held to be a significant variable. Zeuner (1959) is responsible for the dissemination of a model of Pleistocene river behaviour based on continental investigations. This suggests that upstream terraces are likely to represent aggradation under cold conditions, while conversely the downstream sections would more likely record the high sea-levels assumed or known to occur during interglacial (warm) periods. The suggestion that this model fits the Thames has yet to be fully tested, but examination of the London area does not suggest that it is a meeting zone of 'climatic' and 'thalassostatic' terraces. Another problem that has caused a good deal of confusion is the conflicting evidence provided by terrace gravel and terrace flat. A flight of terraces will have terrace surfaces which show a simple correlation of age and

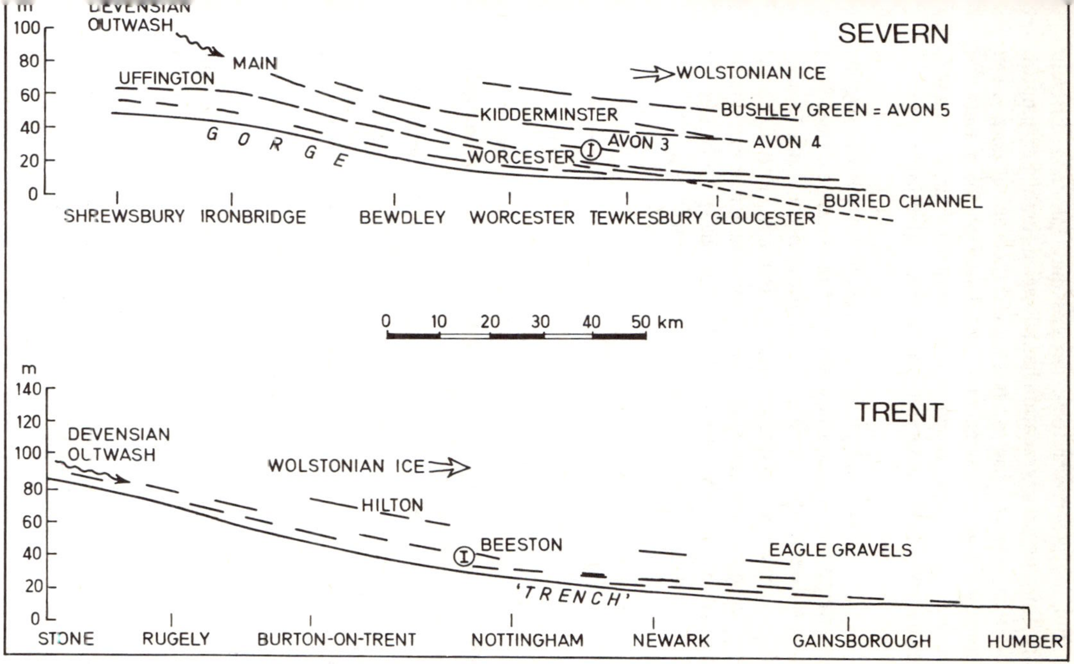

Fig. 14.4 Terrace profiles of the rivers Trent and Severn.

Note the contrast between the Trent terraces which slope at the same angle as the modern floodplain and the Severn terraces which reflect the change in volume after diversion of the Upper Severn through the Ironbridge gorge to form the modern river. The glaciation symbol is located at a height that reflects its stratigraphical position.

height. But the gravels of these terraces may fall in a different sequence if prolonged aggradation has allowed a fill thicker than two terraces to be built up. The top terrace will then be cut on a gravel younger than the lower terrace and ages based on the gravels will conflict with the dates based on the morphological sequence. Different workers will inevitably pay greater attention to one of these lines of evidence.

In practice this model in turn is crude and over-simplified. No one has ever attempted to suggest how a lower flat could be eroded by a meandering river on the old basal gravel of the aggradation without some disturbance and palaeolontological contamination of the old deposit. Yet geological investigation of the fossil material in the gravels rarely considers this possibility. Finally, the whole concept of terrace building by aggradation needs reassessment. Most terrace gravels are not thicker than the gravels beneath the present floodplain of the river, which is determined, not by the aggradation, but by the channel depth in flood. Floodplains and most terraces are cut-and-fill desposits and composed predominantly of point-bar deposits.

The terraces of the Severn and Avon are the best known in the area and illustrate many of the points in the preceding paragraphs. Reference has already been made to the Woolridge terraces which may predate the Wolstonian and here only the later terraces will be considered. The Avon sequence (Tomlinson 1925) begins with No. 5 which is probably contemporaneous with outwash from the retreating Wolstonian ice and the spectacular incision of the infant Avon in the deposits of Lake Harrison. It does not extend very far up the Avon valley and has the rather irregular gradient that has been assumed to be indicative of outwash or fluvioglacial gravels, although it is in fact more likely to be a consequence of its age and the extent of later dissection. The equivalent Bushley Green terrace is even less well-developed in the Stour (Severn) valley and slopes steeply upstream, reflecting the small discharge of the Stour basin before the Severn diversion. Below are three Avon terraces (4, 3 and 2) that are fairly readily correlated with the Severn-Stour sequence (see Table 14.1). The presence of *Hippopotamus* in Avon No. 3 is indicative of an Ipswichian date (Shotton 1953). Avon No. 2 is graded to the same level as the Main terrace of the Severn at Tewkesbury, so their equivalence is established. As we have seen, the Main terrace of the Severn represents the first stage of the Ironbridge diversion with a

Table 14.1 Terraces of the principal rivers

Glacial/ Interglacial	*Severn (Stour)*	*Avon*	*Trent*	*Upper Thames*	*Lower Thames*
Anglian					Winter Hill
Hoxnian					Boyn Hill
Wolstonian	Woolridge	Lake Harrison			
	Bushley Green	No. 5	U. Hilton	Hanborough	Lynch Hill
	Kidderminster	No. 4	L. Hilton	Wolvercote	Taplow
Ipswichian	?Kidderminster	No. 3	Beeston	Summertown-Radley	Upper Floodplain terrace
Devensian	Main	No. 2	Floodplain terrace	Floodplain	Buried channel
Flandrian	Worcester	No. 1	Floodplain (below Beeston lock)		Floodplain

considerable discharge of water and sediment derived partly from the Irish Sea ice-sheet (as is readily seen from the erratic content of the gravels) and from the rapid excavation of a gorge from Ironbridge southwards. In addition to this chronological link, there is the radiocarbon date from Upton Warren. The terrace of the Salwarpe has yielded a considerable fauna and flora that indicates a treeless although by no means a tundra climate. The 14C date of 42000 yr B.P. is considered reliable (Coope, Shotton and Strachan 1961).

The lowest terraces of the Avon and Severn are closely related to the present floodplain, although sections show that the floodplain is cut into the slightly higher (1-3 m) surface of the low No. 1 terrace. On the Severn the correlative Worcester terrace has yielded *Elephas primigenius* and Wills (1924) thought that it predated the buried channel of that river. This channel is about 13 m deep below the floodplain level at Tewkesbury and Wills (1938) reported it as about 32 m deep in the tidal reaches of the Severn. Recent work by Beckinsale and Richardson (1964) suggests that the channel floor is at -8.5 m O.D. at Gloucester, -14 m at Sharpness and -22 m at the Severn tunnel. These two writers described the infill of this channel and attempted to correlate the terrace phases with glacial readvances in the Cheshire Basin. The general relationship is a probable one but it is doubtful if individual terraces can satisfactorily be traced through the Ironbridge gorge section of the Severn, while as we have seen our knowledge of the Cheshire Plain glaciations leaves much to be desired.

The river Trent

The Severn and Avon terraces tend to occur considerably higher above present alluvium than their present-day equivalents in the Trent valley (table 14.1). The highest late Wolstonian terrace of the Trent (the Upper Hilton) is at about 27 m above the floodplain, and the Lower Hilton is at 17-18 m. The terraces are probably the equivalent of the Avon Nos 5 and 4 which are found appreciably higher above the present floodplain due to the rapid incision of the Avon and the post-Main terrace activity of the new river Severn. Palaeontologically the Beeston terrace near Derby (surface about 10 m above alluvium) is the equivalent of Avon No. 3 since both contain *Hippopotamus* (Bemrose and Deeley 1896; Jones and Stanley 1974). The Geological Survey (Mitchell and Stevenson 1955) has ignored the subdivision of the Hilton terrace and by joining up all the heights on all the terrace flats have tried to suggest it is fluvioglacial outwash. The mechanics of sedimentation to give such an irregular surface were not explained. Posnansky (1960) has suggested that the uppermost gravels are so coarse that they represent an outwash stream, augmented by melting ice. He also noted that the section at Willington (Lower Hilton) showed cryoturbated gravels below with gravels free of such features above. This is strongly reminiscent of the No. 4 terrace of the Avon. Ice retreat will have occurred a little later in the Trent valley than in the Avon, but in general it seems likely that the Upper Hilton terrace is the general equivalent of No. 5 terrace, and that, like it, it does include direct glacial outwash.

As Posnansky (1960) pointed out, this correlation of the terraces is supported by the archaeological evidence. The Hilton gravels contain implements ranging up to Levelloisian and late Acheulian, while the Beeston terrace has yielded only fresh Levallois material. The low terrace of the Trent (much of it below the floodplain above Beeston Lock) has yielded a cold fauna, including *Elephas primigenius* and *Tichorhinus antiquitatus*.

The terraces described here pass down the present line of the Trent valley to Newark and then to Lincoln, although the constricted trench-like character of the stretch between Nottingham and Newark has not allowed either of the Hilton terraces to be preserved. On the basis of longitudinal slopes Clayton (1953b) correlated the Upper Hilton with the Eagle gravels between Newark and Lincoln but whether those spreads are related to sea-level or to glacial ponding, as Straw (1963) suggests, is obscure. For a fuller discussion see pp.30-1 in this volume.

The Thames

The classic work of Sandford and Arkell in particular established a sequence of terraces in the Thames valley above Goring gap. The highest widespread terrace is the Wolvercote terrace which is at about 16 m above the river at Oxford. Bishop (1957) has shown that this correlates with the overflow through the Fenny Compton gap and down the Cherwell from Lake Harrison. It includes frequent flints derived from the Wolstonian outwash. Below the Wolvercote is the Summerton-Radley terrace at about 8 m above the present floodplain. The upper part of this terrace is warm, and of Ipswichian age. The correlation is pretty firm, for the lower floodplain gravels (which project above the present floodplain surface in places) have a thoroughly cold fauna and flora and must represent the Devensian. There are scattered relics of older and higher terraces, notably the Hanborough (32 m above alluvium), once thought to be the upstream continuation of the Boyn Hill (largely on grounds of height). Recently Briggs and Gilbertson (1973) have shown that the sedimentology and fauna indicate a cold climate, so it, too, must be Wolstonian. It thus seems that the Hanborough must be the upstream equivalent of the Lynch Hill terrace, the Wolvercote with the Taplow, and the Summertown-Radley with the Upper Floodplain terrace. In all three cases this gives a downvalley gradient of about 0.4 m per km, close to that of the modern floodplain. Since the gravels have been reworked at each stage, this consistent gradient supports the correlations suggested here.

Upland peat deposits

The most recent event of significance has been the postglacial accumulation of peat on the wetter uplands of the Pennines, and most recently its dissection. Wide areas in the southern Pennines above 500 m are covered with thick peat, often of the order of 4 m or more. Quite large areas are dissected by gullies, in places leaving isolated mounds of peat known locally as haggs. We have some idea of the chronology of peat accumulation and erosion, but the causes are the subject of controversy. Moss (1913) described both the peat and its dissection, noting the headward extension of the gullies shown by comparing separate editions of the Ordnance Survey 1:50000 map. Conway, first on Ringinglow Bog (1947) and later across a much larger part of the southern Pennines (1954) established through pollen analysis that the peat began to accumulate around 8000 B.P., at the transition from Boreal to Atlantic times. Accumulation was relatively slow for about 2500 years, but has generally been much faster later. The broadest and hence least well drained peat areas have developed thick and poorly consolidated *Sphagnum* peats: these are thickest at at highest elevations where precipitation is greatest and evapotranspiration the lowest. Peat accumulation is linked with a great reduction in forest cover. It is not established whether this is solely related to climatic change or whether man played a role (perhaps dominant).

Conway (1954) linked the *Sphagnum* peat with the tendency for erosion, suggesting that this weak material was liable to failure, especially on slopes. Once started, headward gully extension would dissect the peat over wide areas. While such gullies tended to become revegetated in Scotland, the greater disturbance by fire (and also trampling?) in the Pennines allowed erosion to continue until all the peat was removed. The rate of erosion is a function of slope and runoff, hence the margins of the highest plateaux are worst affected. Bower (1960, 1961) mapped peat erosion over the whole of the Pennines, dividing it into several categories, largely on the basis of gully pattern and extent of erosion. Like Conway she regarded erosion as an inevitable stage in the thickening of peat and primarily a function of climate. All other effects were thought to be secondary, although it was agreed that fire, man, grazing and air pollution all stressed the environment and made erosion more likely and more widespread. In contrast Radley (1962) drew attention to the recency of widespread peat erosion, and concluded that human action needs much greater emphasis. He also felt that wind played an important role in the widening of gullies and the recession of marginal faces around peat-covered plateaux. His emphasis on the speed of erosion in recent years seems justified, and this suggests that human pressure has triggered erosion, but its broad distribution is also a function of climate.

References

Arkell, W.J. (1947), *The geology of Oxford.* Oxford.

Armstrong, A.L. (1931), Excavations in the Pin Hole Cave, Creswell Crags, Derbyshire. *Proc. Prehist. Soc. E. Anglia.* 6, 330-4.

Armstrong, A.L. (1939), Palaeolithic Man in the North Midlands. *Mem. & Proc. Manchester Lit. & Phil. Soc.* 83, 87-116.

Beckinsale, R.P. and Richardson, L. (1964), Recent findings on the physical development of the Lower Severn Valley. *Geogr. J.* 130, 87-105.

Bemrose, H.H. and Deeley, R.M. (1896), Discovery of mammalian remains in the Old River Gravels of the Derwent near Derby. *Q. J. Geol. Soc. Lond.* 52, 497-510.

Bird, J.B. (1967), *The physiography of Arctic Canada, with special reference to the area south of Parry Channel.* Baltimore.

Bishop, W.W. (1957), The Pleistocene geology and geomorphology of three gaps in the Midland Jurassic scarp. *Phil. Trans. R. Soc. B.* 241, 255-305.

Boulter, M.C., Ford, T.D., Ijtaba, M. and Walsh, P.T. (1971), The Brassington Formation: a newly recognized Tertiary formation in the southern Pennines. *Nature: Physical Science.* 231, 134-6.

Boulton, G.S., Jones, A.S., Clayton, K.M. and Kenning, M.J. (1977), A British ice-sheet model and patterns of glacial erosion and deposition in Britain. In *British Quaternary Studies - recent advances.* Oxford, 231-46.

Boulton, G.S. and Worsley, P. (1965), Late Weichselian glaciation in the Cheshire-Shropshire basin. *Nature.* 207, 704-6.

Bower, M.M. (1960), The erosion of blanket peat in the southern Pennines *E. Midl. Geogr.* 13, 22-33.

Bower, M.M. (1961), The distribution of erosion in blanket peat bogs in the Pennines. *Trans. Inst. of Brit. Geog.*, 29, 17-30.

Briggs, D.J. and Gilbertson, D.D. (1973), The age of the Hanborough terrace of the river Evenlode, Oxfordshire. *Proc. Geol. Assoc.* 84, 155-73.

Bunting, B.T. (1961), The role of seepage moisture in soil formation, slope development, and stream initiation. *Am. J. Sc.* 259, 503-18.

Chandler, R.J. (1970), The degradation of Lias Clay slopes in an area of the East Midlands. *Q. J. Eng. Geol.* 2, 161-81.

Chandler, R.J. (1971), Landsliding on the Jurassic escarpment near Rockingham, Northamptonshire. In *Slopes, form and process* (Institute of British Geographers, special publication No.3), 111-28.

Clayton, K.M. (1953a), The denudation chronology of part of the middle Trent basin. *Trans. Inst. of Brit. Geogr.* 19, 25-36.

Clayton, K.M. (1953b), The glacial chronology of part of the middle Trent basin. *Proc. Geol. Assoc.* 64, 198-207.

Clayton, K.M. (1955), The geomorphology of the area around Nottingham and Derby. *E. Midl. Geogr.* 3, 16-20.

Clayton, K.M. (1957), The differentiation of the glacial drifts of the East Midlands. *E. Midl. Geogr.* 7, 31-40.

Clayton, K.M. (1959), The geomorphology of the area around Melton Mowbray. *E. Midl. Geogr.* 12, 3-8.

Clayton, K.M. (1968), Structure-surface relationships in the middle part of the Derwent Basin. *E. Midl. Geogr.* 4, 321-8.

Clayton, K.M. (1977), River terraces. In *British Quaternary Studies-recent advances*, Oxford, 153-67.

Conway, V.M. (1947), Ringinglow Bog, near Sheffield. *J. of Ecology*, 34, 149-81.

Conway, V.M. (1954), Stratigraphy and pollen analysis of southern Pennine blanket peats. *J. Ecol.* 42, 117-47.

Coope, G.R., Shotton, F.W. and Strachan, I. (1961), A Late Pleistocene fauna and flora from Upton Warren, Worcestershire. *Phil. Trans. R. Soc. B*, 244, 379-421.

Dalton, A.C. (1958), The distribution of dolerite boulders in the glaciation of N. E. Derbyshire. *Proc. Geol. Assoc.* 68, 278-85.

Dearden, J. (1963), Derbyshire limestone, its removal by man and nature. *E. Midl. Geogr.* 3, 199-205.

Dobson, M.R. (1977a), The geological structure of the Irish Sea. In C. Kidson and M.J. Tooley (eds) *The Quaternary history of the Irish Sea.* Liverpool, 13-26.

Dobson, M.R. (1977b), The history of the Irish Sea basins. In C. Kidson and M.J. Tooley (eds) *The Quaternary history of the Irish Sea.* Liverpool, 93-8.

Dunham, K.C. (1973), A recent deep borehole near Eyam, Derbyshire. *Nature: Physical Science* 241, 84-5.

Dury, G.H. (1951a), A 400-foot bench in south-eastern Warwickshire. *Proc. Geol. Assoc.* 62, 167-73.

Dury, G.H. (1951b), The summit-plane of the Jurassic hills northeast of the Cotswolds. *J. Northants. Nat. Hist. Soc. and Field Club* 32, 55-60.

Early, K.R. and Skempton, A.W. (1972), Investigations of the landslide at Walton's Wood, Staffordshire. *Q. J. Eng. Geol.* 5, 19-41.

Earp, J.R. and Hains, B.A. (1971), *The Welsh Borderland* 3rd edn. (British Regional Geology). London.

Earp, J.R., Magraw, D., Poole, E.G., Land, D.H. and Whiteman, A.J. (1961), *Geology of the country around Clitheroe and Nelson.* Memoirs Geological Survey Great Britain, sheet 68.

Eastwood, T., Gibson, W., Cantrill, R.C. and Whitehead, T.H. (1923), *The geology of the country around Coventry, including an account of the Carboniferous rocks of the Warwickshire Coalfield.* Memoirs Geological Survey Great Britain, sheet 169.

Edwards, E.A. and Dinham, C.H. (1965) *Geology of the country around Huntingdon and Biggleswade.* Memoirs Geological Survey Great Britain, sheets 187 and 204.

Fearnsides, W.G. (1932), The geology of the eastern part of the Peak District. *Proc. Geol. Assoc.* 43, 152-91.

Ford, T.D. (1967), Deep weathering, glaciation and tor formation in Charnwood Forest, Leicestershire. *Mercian Geol.* 2, 3-14.

Ford, T.D. (1968), The morphology of Charnwood Forest. In P.C. Sylvester-Bradley and T.D. Ford (eds), *The geology of the East Midlands.* Leicester.

Ford, T.D. (ed.) (1977), *Limestones and caves of the Peak District.* Norwich.

Franks, J.W. and Johnson, R.H. (1964), Pollen analytical dating of a Derbyshire landslip: the Cown Edge landslides, Charlesworth. *New Phytologist* 63, 209-16.

George, T.N. (1974), Prologue to a geomorphology of Britain. In E.H. Brown and R.S. Waters (eds) *Progress in geomorphology: papers in honour of D.L. Linton*. Institute of British Geographers, special publication 7, 113-25.

Greig, D.C., Wright, J.E., Hains, B.A. and Mitchell, G.H. (1968), *Geology of the country around Church Stretton, Craven Arms, Wenlock Edge and Brown Clee*. Memoirs Geological Survey of Great Britain, sheet 166.

Gresswell, R.K. (1964), The origin of the Mersey and Dee estuaries. *Geol. J.* 4, 77-86.

Hains, B.A. and Horton, A. (1969), *Central England* 3rd edn. (British Regional Geology). London.

Harmer, F.W. (1928), On the distribution of erratics and drift. *Proc. Yorks. Geol. Polytech. Soc.* 21, 79-150.

Hey, R.W. (1958), High-level gravels in and near the Lower Severn Valley. *Geol. Mag.* 95, 161-8.

Hey, R.W. (1959), Pleistocene deposits on the west side of the Malvern Hills. *Geol. Mag.* 96, 403-17.

Hollingworth, S.E. and Kellaway, G.A. (1944), Large-scale superficial structures in the Northamptonshire Ironstone field. *Q. J. Geol. Soc. Lond.* 100, 1-44.

Hollingworth, S.E. and Taylor, J.H. (1946), An outline of the geology of the Kettering district. *Proc. Geol. Assoc.* 62.

Johnson, R.H. (1957-8), Observations on the stream patterns of some peat moorlands in the southern Pennines. *Mem. and Proc. Manchester Lit. and Phil. Soc.* 49, 1-18.

Johnson, R.H. (1965a), The glacial geomorphology of the west Pennine slopes, from Congleton to Cliviger. In J.B. Whittow and P.D. Wood (eds), *Essays in Geography for Austin Miller*. Reading. 58-93.

Johnson, R.H. (1965b), The origin of the Churnet and Rudyard valleys. *N. Staffs. J. Field Studs.* 5, 95-105.

Johnson, R.H. (1965c), A study of the Charlesworth landslides near Glossop, north Derbyshire. *Trans. Inst. Brit. Geogr.* 37, 111-26.

Johnson, R.H. and Rice, R.J. (1961), Denudation chronology of the southwest Pennine Upland. *Proc. Geol. Assoc.* 72, 21-31.

Jones, P.F. and Stanley, M.F. (1974), Ipswichian mammalian fauna from the Beeston terrace at Boulton Moor, near Derby. *Geol. Mag.* 111, 515-20.

Jowett, A. and Charlesworth, J.K. (1929), The glacial geology of the Derbyshire Dome and the western slopes of the southern Pennines. *Q. J. Geol. Soc. Lond.* 85, 307-34.

Jukes-Browne, A.J. (1887), *The geology of part of East Lincolnshire* Memoirs Geological Survey.

Kellaway, G.A. and Taylor, J.H. (1953), Early stages in the physiographic evolution of a portion of the East Midlands. *Q. J. Geol. Soc. Lond.* 108, 343-76.

Kent, P.E. (1949), A structure contour map of the surface of the buried pre-Permian rocks of England and Wales. *Proc. Geol. Assoc.* 60, 87-104.

Kent, P.E. (1967), A contour map of the sub-Carboniferous surface in the north-east Midlands. *Proc. Yorks. Geol. Soc.* 36, 127-33.

King, C.A.M. (1960), The Churnet Valley. *E. Midl. Geogr.* 14, 33-40.

King, C.A.M. (1962), Geomorphological contrasts in New Zealand and Britain. In M. McCaskill (ed.) *Land and livelihood.* Christchurch, New Zealand, 48-73.

King, C.A.M. (1976, *Northern England.* London.

King, C.A.M. (1977), The early Quaternary landscape with consideration of neotectonic matters. In *The Quaternary of the British Isles - recent advances.* Oxford, 137-52.

Lamplugh, G.W. *et al* (1909), *The geology of the Melton Mowbray district and south-east Nottinghamshire.* Memoirs Geological Survey England and Wales, sheet 142.

Linton, D.L. (1951a), Midland drainage: some considerations bearing on its origin. *Adv. Sci.* 7, 449-56.

Linton, D.L. (1951b), Problems of Scottish scenery. *Scot. Geog. Mag.* 67, 65-85.

Linton, D.L. (1956), Geomorphology. In D.L. Linton (ed.), *Sheffield and its region.* British Association for the Advancement of Science.

Linton, D.L. (1964), The origin of the Pennine tors — an essay in analysis. *Zeit. für Geomorph.* 8, *Sonderheft*, 5*-24*.

McArthur, J.L. (1977), Quaternary erosion in the upper Derwent basin and its bearing on the age of surface features in the southern Pennines. *Trans. Inst. Brit. Geogr.* 2, 490-7.

Mitchell, G.H. and Stevenson, I.P. (1955), *Geology of the country between Burton-on-Trent, Rugeley and Uttoxeter.* Memoirs Geological Survey, U.K., sheet 140.

Morgan, A.V. (1973), The Pleistocene geology of the area north and west of Wolverhampton, Staffordshire, England. *Phil. Trans. R. Soc. Lond.* 265, 233-97.

Moss, C.E. (1913), *The vegetation of the Peak District*. Cambridge.

Ollier, C.D. and Thomasson, A.J. (1957), Asymmetrical valleys of the Chiltern Hills. *Geogr. J.* 123, 71-80.

Palmer, J. and Radley, J. (1961), Gritstone tors of the English Pennines. *Zeit. für Geomorph.* 5, 37-52.

Pickering, R. (1957), The Pleistocene geology of the south Birmingham area. *Q. J. Geol. Soc. Lond.* 113, 223-37.

Pigott, C.D. (1962), Soil formation and development on the Carboniferous limestone of Derbyshire. Part 1, parent materials. *J. Ecol.* 50, 145-56.

Pigott, C.D. (1965), Limestone surfaces under superficial deposits in north Derbyshire. *Geogr. J.* 131, 41-4.

Pitty, A.F. (1965), A study of some escarpment gaps in the southern Pennines. *Trans. Inst. Brit. Geogr.* 37, 127-45.

Pitty, A.F. (1967), Some problems in the relation between geology and relief in the southern part of the Ashover dome, Derbyshire. *E. Midl. Geogr.* 4, 176-84.

Pitty, A.F. (1968a), The scale and significance of solutional loss from the limestone tract of the southern Pennines. *Proc. Geol. Assoc.* 79, 153-78.

Pitty, A.F. (1968b), *An approach to the study of karst waters*. University of Hull, Occasional Papers, 5.

Pocock, R.W., Whitehead, T.H., Wedd, C.B. and Robertson, T. (1938), *Shrewsbury District including the Hanwood Coalfield*. Memoirs Geological Survey Great Britain, sheet 152.

Poole, E.G. and Whiteman, A.J. (1961), The glacial drifts of the southern part of the Shropshire-Cheshire basin. *Q. J. Geol. Soc. London.* 117, 91-130.

Poole, E.G. and Whiteman, A.J. (1966), *Geology of the country around Nantwich and Whitchurch*. Memoirs Geological Survey Great Britain, sheet 122.

Posnansky, M. (1960), The Pleistocene succession in the middle Trent basin. *Proc. Geol. Assoc.* 71, 285-311.

Radley, J. (1962), Peat erosion on the high moors of Derbyshire and West Yorkshire. *E. Midl. Geogr.* 3, 40-50.

Rice, R.J. (1968), The Quaternary deposits of central Leicestershire. *Phil. Trans. R. Soc. Lond.* 262, 459-509.

Rieuwerts, J.H. (1966), The soughs of the Derbyshire lead mines. *Bull. Peak District Mines Hist. Soc.* 3, 1-42.

Rieuwerts, J.H. (1969), The soughs of the Derbyshire lead mines — supplementary list. *Bull. Peak District Mines Hist. Soc.* 4, 119-35.

Sandford, K.S. (1929), The erratic rocks and the southern limit of glaciation in the Oxford district. *Q. J. Geol. Soc. Lond.* 85.

Shaw, J. (1972), Sedimentation in the ice-contact environment, with examples from Shropshire (England). *Sedimentology.* 18, 23-62.

Shotton, F.W. (1953), The Pleistocene deposits of the area between Coventry, Rugby and Leamington and their bearing on the topographic development of the Midlands. *Phil. Trans. R. Soc. B.* 237, 209-60.

Shotton, F.W. (1960), Large-scale patterned ground in the valley of the Worcestershire Avon. *Geol. Mag.* 97, 404-8.

Shotton, F.W. (1967), The age of the Irish Sea glaciation of the Midlands. *Nature.* 215, 1366.

Shotton, F.W. (1977), *The English Midlands.* (Tenth I.N.Q.U.A. Field Guide, A2).

Shotton, F.W., Banham, P.H. and Bishop, W.W. (1977), Glacial-interglacial stratigraphy of the Quaternary in Midland and eastern England. In *British Quaternary Studies - recent advances.* Oxford. 267-82.

Shotton, F.W. and Wilcockson, W.H. (1951), Superficial valley folds in an opencast working of the Barnsley coal. *Proc. Yorks. Geol. Soc.* 28, 102-11.

Simpson, I.M. (1959), The Pleistocene succession in the Stockport and South Manchester area. *Q. J. Geol. Soc. Lond.* 115, 107-21.

Simpson, I.M. and West, R.G. (1958), On the stratigraphy and palaeobotany of a Late-Pleistocene organic deposit at Chelford, Cheshire. *New Phytologist.* 57, 239-50.

Sissons, J.B. (1954), The erosion surfaces and drainage systems of south-west Yorkshire. *Proc. Yorks. Geol. Soc.* 29, 305-42.

Stephens, J.V. Mitchell, G.B. and Edwards, W. (1953), *Geology of the country between Bradford and Skipton*, Memoirs Geological Survey Great Britain, sheet 69.

Straw, A. (1963), The Quaternary evolution of the lower and middle Trent. *E. Midl. Geogr.* 3, 171-89.

Straw, A. and Lewis, G.M. (1962), Glacial drift in the area around Bakewell, Derbyshire. *E. Midl. Geogr.* 3, 72-80.

Sweeting, G.S. (1946), An outline of the geology of Ashover, Derbyshire, *Proc. Geol. Assoc.* 57, 117-36.

Sweeting, M.M. (1965), Denudation in limestone regions, *Geog. J.* 131, 34-7.

Swinnerton, H.H. (1937), The geology of the Nottingham district. In: *Scientific survey of Nottingham and district*. British Association Advancement Science, 45.

Sylvester-Bradley, P.C. and Ford, T.D. (1968), *The geology of the East Midlands*, Leicester.

Tomlinson, M.E. (1925), River terraces of the lower valley of the Warwickshire Avon. *Q. J. Geol. Soc. Lond.* 81, 137-68.

Tomlinson, M.E. (1941), Pleistocene gravels of the Cotswold sub-edge plain from Mickleton to the Frome valley. *Q. J. Geol. Soc. Lond.* 96, 385.

Walsh, P.T., Boulter, M.C., Ijtaba, M. and Urbani, D.M. (1972), The preservation of the Neogene Brassington Formation of the southern Pennines and its bearing on the evolution of upland Britain. *J. Geol. Soc.* 128, 519-59.

Warwick, G.T. (1963), The dry valleys of the southern Pennines, England. *Erdkünde.* 18, 116-22.

Waters, R.S. and Johnson, J.H. (1958), The terraces of the Derbyshire Derwent. *E. Midl. Geogr.* 9, 3-15.

Watts, W.W. (1903), A buried Triassic landscape. *Geogr. J.* 21, 623-36.

Watts, W.W. (1947), *Geology of the ancient rocks of Charnwood Forest*. London.

Whitehead, T.H., Eastwood, T. and Robertson, T. (1927), *Geology of the southern part of the South Staffordshire coalfield.* Memoirs Geological Survey Great Britain.

Whitehead, T.H. and Pocock, R.W. (1947), *Dudley and Bridgnorth.* Memoirs Geological Survey Great Britain, sheet 167.

Wills, L.J. (1924), The development of the Severn valley in the neighbourhood of Iron-Bridge and Bridgnorth. *Q. J. Geol. Soc. Lond.* 80, 274-314.

Wills, L.J. (1937), The Pleistocene history of the West Midlands. *Adv. Sci.* 71-94.

Wills, L.J. (1938), The Pleistocene development of the Severn from Bridgnorth to the sea. *Q. J. Geol. Soc. Lond.* 94, 161-242.

Wills, L.J. (1950), *The palaeogeography of the Midlands.* 2nd edn, London.

Wills, L.J. (1970), The Triassic succession in the central Midlands in its regional setting. *Q. J. Geol. Soc. Lond.* 126, 225-85.

Wolfenden, E.B. (1958), Palaeoecology of the Carboniferous reef complex and shelf limestones in northwest Derbyshire. *Bull. Geol. Soc. America.* 69, 871-98.

Wooldridge, S.W. (1938), The glaciation of the London Basin and evolution of the lower Thames drainage system. *Q. J. Geol. Soc. Lond.* 94, 627-67.

Wooldridge, S.W. and Linton, D.L. (1955), *Structure, surface and drainage in south-east England.* 2nd edn London.

Worsley, P. (1966), Some Weichselian fossil frost wedges from east Cheshire. *Merc. Geol.* 1, 357-65.

Worsley, P. (1967), Problems in naming the Pleistocene deposits of the north-east Cheshire Plain. *Merc. Geol.* 2, 51-5.

Worsley, P. (1970), The Cheshire-Shropshire lowlands. In C.A. Lewis (ed.) *The glaciations of Wales and adjoining regions.* London, 83-106.

Worsley, P. (1975), An appraisal of the glacial Lake Lapworth concept. In A.D.M. Phillips and B.J. Turton (eds) *Environment, man and economic change.* London, 98-118.

Wright, W.B. (1914), *The Quaternary Ice Age.* 1st edn (2nd edn 1936) London.

Yates, E.M. (1956), The Keele surface and the Upper Trent drainage *E. Midl. Geogr.* 5, 10-21.

Yates, E.M. and Moseley, F. (1958), Glacial lakes and spillways in the vicinity of Madeley, north Staffordshire. *Q. J. Geol. Soc. Lond.* 113, 409-28.

Young, A. (1974), The rate of slope retreat. In *Progress in geomorphology.* Institute British Geographers, special publication 7, 65-78.

Zeuner, F.E. (1959), *The Pleistocene period: its climate, chronology and faunal successions.* 2nd edn, London.

Index